Owens Perceptual Multiverse Theory

Anthony Scott Owens

Preface to the Owens Perceptual Multiverse Theory

The Owens Perceptual Multiverse Theory is the culmination of a deep exploration into the nature of reality, human perception, and the intricacies of how individuals and societies construct their worlds. Rooted in the belief that reality is not a singular, monolithic experience, this theory posits that each of us lives within a multiverse of thought—an ever-expanding network of parallel realities shaped by our beliefs, experiences, and perceptions. This is not just a metaphor for how people think; it's a literal model for understanding how our minds and societies construct different dimensions of reality.

The Genesis of the Theory

In formulating the Owens Perceptual Multiverse Theory, my starting point was a simple but profound observation: two people can witness the same event, live through the same circumstances, and yet come away with radically different interpretations of what is "real." This realization led me to question whether reality itself was objective, or if it was far more fluid, determined by the filters—cognitive, cultural, and experiential—through which each individual perceives the world.

In my experiences managing disasters, leading people through crises, and advocating for social and political change, I observed time and again that individuals facing the same external events often inhabit vastly different worlds. These are not mere differences in opinion; they are reflections of the dimensions of reality people occupy based on how they perceive and interact with the world. From here, the idea of a Perceptual Multiverse emerged, where each dimension is not a distant, hypothetical parallel universe but rather a subjective reality that each of us navigates every day.

The Core Premise

The Owens Perceptual Multiverse Theory is built on the foundational idea that human beings do not live in a single, objective reality but rather in a multiverse of perceptions. Each person's beliefs, experiences, thoughts, and emotions form the boundaries of their personal dimension within this multiverse. These dimensions are fluid and dynamic, meaning that individuals can shift between realities based on new experiences, changing perceptions, or external crises. At certain points—what I call flashpoint issues—these realities can converge, temporarily collapsing parallel dimensions into shared experiences that bring people together, only to fragment again as individuals return to their personal perceptions.

This theory suggests that reality is subjective, influenced by personal experience, cultural narratives, and belief systems, but it also operates within the broader collective consciousness. Individuals shape their realities, but these realities are also shaped by the dimensions of thought that we inhabit as societies. The result is a multiverse where dimensions exist not just on a physical plane but on a cognitive, emotional, and societal level.

The Importance of Dimensional Awareness

One of the most powerful implications of the Owens Perceptual Multiverse Theory is the notion of dimensional awareness—the ability to recognize that the reality we inhabit is but one of many. By becoming aware of the dimensions we live in, we gain the power to navigate between them, understand others' realities, and expand our consciousness to encompass broader perspectives. Empathy, critical thinking, and open-mindedness are crucial tools in this process, as they allow us to step outside of our own dimensions and engage with the realities of others.

This awareness can transform how we approach conflict, both on a personal and societal level. When we understand that others are operating from different dimensions of perception, it becomes easier to engage in dialogue, find common ground, and collaborate toward solutions that benefit multiple realities. Dimensional shifts are not just theoretical—they have real-world implications for politics, social justice, cultural understanding, and human connection.

Flashpoint Issues: The Gateways to New Realities

A major component of this theory is the idea of flashpoint issues—societal or personal moments of intense conflict or transformation that serve as gateways to new dimensions of thought. These flashpoints are moments when different dimensions collide, creating opportunities for dimensional expansion or contraction. Whether it's a political debate, a natural disaster, or a cultural movement, flashpoint issues force individuals to confront the boundaries of their reality, often leading to profound shifts in perception.

Flashpoint issues are critical because they challenge the status quo and serve as catalysts for growth. When confronted with these intense moments of dimensional friction, individuals and societies must either evolve their perceptions or become entrenched in their current dimensions. In either case, the flashpoint serves as a pivotal moment in the evolution of the Perceptual Multiverse.

Disasters and Dimensional Convergence

Disasters, in particular, play a unique role in the Owens Perceptual Multiverse Theory as moments of dimensional synchronization. During crises, whether personal or collective, individuals who typically inhabit disparate realities often find themselves temporarily aligned within a shared dimension. This phenomenon occurs because the external crisis forces people to prioritize the same concerns—survival, safety, and recovery—over their usual subjective experiences. Disasters are potent dimensional gateways, collapsing the multiverse into unified experiences that transcend individual differences, if only for a moment. However, once the crisis abates, people often retreat back into their personal dimensions, leading to the fragmentation of the shared reality.

Understanding how these moments of synchronization occur allows us to better navigate global challenges, from natural disasters to pandemics to political upheavals. By recognizing the dimensional fluidity that crises create, we can harness these moments of shared reality to foster cooperation, empathy, and collective action.

The Future of the Owens Perceptual Multiverse Theory

As I continue to develop and refine this theory, I envision its applications extending far beyond the realms of philosophy and sociology. The Owens Perceptual Multiverse Theory offers a lens through which to understand not only human thought and behavior but also the future of technology, artificial intelligence, and cognitive science. As we become more aware of the multidimensional nature of reality, the potential for personal and collective evolution becomes limitless.

This theory challenges us to move beyond the narrow confines of our individual realities and embrace the infinite possibilities that exist within the multiverse of thought. It invites us to recognize that reality is fluid, that we are constantly shifting between dimensions, and that our perception of the world is a powerful force in shaping both our personal experience and the collective future of humanity.

A Personal Journey

The Owens Perceptual Multiverse Theory is deeply personal for me. It is born out of my life's experiences—my roles as a father, husband, farmer, emergency manager, entrepreneur, and thinker. Through these diverse roles, I have witnessed firsthand how people inhabit different realities, often without realizing it, and how this lack of awareness can lead to misunderstandings, conflict, and missed opportunities for growth.

In developing this theory, my hope is to offer a new way of understanding not only the world around us but also ourselves. By embracing the idea of the Perceptual Multiverse, we can learn to move beyond the limitations of our own dimensions, connect with others in more meaningful ways, and co-create a future that is shaped by compassion, curiosity, and collective wisdom.

This theory is not just an intellectual exercise—it is a call to action for anyone who seeks to understand the deeper layers of reality and contribute to the evolution of human consciousness.

Table of Contents

Introduction: The Perceptual Multiverse Unveiled

- Defining the Multiverse of Thought and Belief
- The Nature of Reality: Objective vs. Subjective Worlds
- The Role of Perception in Shaping Dimensions

Chapter 1: Foundations of the Multiverse

- Scientific Roots of the Multiverse Theory
- From Quantum Physics to Perceptual Realities
- The Observer Effect and Reality Creation

Chapter 2: Dimensions Beyond Perception

- What Are Extra Dimensions?

- Invisible Realities and the Limits of Human Perception
- Unlocking the Potential of Future Dimensions

Chapter 3: Flashpoint Issues as Dimensional Evidence

- Defining Flashpoint Issues: Clashes of Parallel Realities
- Why Flashpoints Are Proof of Different Thought Dimensions
- How Dimensions Overlap and Intersect in Society

Chapter 4: The Environmental Dimension

- Climate Change: The Battle of Two Realities
- Experiencing the Environmental Crisis in Different Dimensions
- The Future of Climate Dimensions: Collapse or Recovery?

Chapter 5: The Health Crisis Dimension

- The Pandemic Divide: Health Science vs. Conspiracy
- Vaccines, Misinformation, and Parallel Perceptions of Risk
- How Flashpoints Shape the Health Crisis Reality

Chapter 6: Gender and Identity in the Multiverse

- Fluid Dimensions: Gender as a Social Construct
- Fixed Dimensions: Binary Gender and Biological Determinism
- The Intersection of Dimensions: Where Identity Becomes Reality

Chapter 7: Racial Inequality and Parallel Histories

- Systemic Racism: A Dimension Rooted in History and Experience
- Meritocracy and Individualism: A Separate Dimension of Opportunity
- Dimensional Collisions: The Struggle for Recognition Across Racial Realities

Chapter 8: The Sanctity of Life Dimension

- Abortion and the Dividing Line of Reality
- Bodily Autonomy vs. The Right to Life: Two Parallel Universes
- The Moral Battle Between Dimensions of Belief

Chapter 9: Economic Inequality as Dimensional Conflict

- Redistribution vs. Free Market: Two Economic Realities
- Capitalism and Socialism as Competing Dimensions
- How Wealth and Power Shape the Dimensional Divide

Chapter 10: The Political Dimensions of Nationalism vs. Globalism

- National Sovereignty and the Battle for Dimensional Control
- Globalization: Connecting or Fragmenting Realities?
- How Nations Create Their Own Dimensions of Perception

Chapter 11: Digital Dimensions and Information Bubbles

- Social Media as the Gateway to Infinite Dimensions
- How Algorithms Create Personalized Realities

- The Digital Multiverse: Echo Chambers, Misinformation, and Reality Distortion

Chapter 12: The Role of Science and Technology in Shaping Dimensions

- Technological Utopias and Dystopias: A Future Reality Split
- Artificial Intelligence: The Catalyst for a New Dimension
- The Ethics of Dimensional Expansion Through Technology

Chapter 13: Crossing the Dimensional Boundaries

- How Do We Shift from One Dimension to Another?
- Cognitive Dissonance as a Symptom of Dimensional Conflict
- Can We Find Common Ground Between Parallel Realities?

Chapter 14: Extra Dimensions Beyond the Flashpoints

- **The Hidden Dimensions Beyond Physical Perception**
    - The concept of dimensions beyond our immediate senses
    - The limitations of the human mind in perceiving the full scope of reality
    - Scientific theories on dimensions we cannot observe (e.g., string theory's curled-up dimensions)
- **Vibrational Dimensions: How Frequency Shapes Reality**
    - The idea that different dimensions vibrate at different frequencies
    - How altering our personal vibrational frequency can allow us to access new realities
    - Techniques for tuning into different vibrational states (e.g., sound therapy, meditation, energy work)
- **Mental Dimensions: The Power of Consciousness**
    - How the mind acts as a portal to other dimensions
    - States of consciousness (meditation, lucid dreaming, and altered states) as gateways to new dimensions
    - The role of imagination and thought in shaping personal dimensions of reality
- **Spiritual Dimensions: Connecting with the Higher Self and Universal Energy**
    - The spiritual belief in higher planes of existence (astral planes, etheric planes)
    - Exploring ancient traditions (shamanic journeys, Tibetan meditation practices, etc.) that access these dimensions
    - Mystical experiences and visions as interactions with spiritual dimensions
- **Vibrational Beings and Entities in Other Dimensions**
    - The possibility of life forms or conscious entities in higher vibrational dimensions
    - Interactions with these beings through spiritual practices, rituals, or altered states
    - Accounts of encounters from various spiritual and cultural perspectives
- **The Akashic Records: A Dimension of Universal Knowledge**
    - Exploring the concept of the Akashic Records, a metaphysical dimension of all knowledge
    - Accessing this dimension through meditative or spiritual practice
    - The role of the Akashic dimension in shaping personal and collective reality
- **Techniques for Accessing Hidden Dimensions**
    - Meditation and breathwork for altering states of consciousness

- Sound and vibration therapy (e.g., Tibetan singing bowls, binaural beats) for dimensional travel
            - Rituals, visualization, and intentional focus to tap into spiritual or vibrational dimensions
            - Technology's emerging role in helping humans access different states of mind and vibration
    - **What Lies Beyond: Unexplored Dimensions Waiting for Human Perception**
        - The potential for new discoveries in vibrational science, consciousness, and spiritual realms
        - Future technologies that may allow us to explore and interact with these dimensions more fully
        - Humanity's journey toward expanding perception and tapping into the full multiverse of existence

Chapter 15: The Perceptual Multiverse and the Human Experience

- How Perception Shapes Reality in the Multiverse
- The Multiverse as a Framework for Understanding Conflict and Cooperation
- Living in Parallel Realities: Embracing the Multiverse in Everyday Life

Chapter 16: Disasters and Dimensional Synchronization

16.1 The Nature of Dimensional Convergence During Disasters

- How External Events Act as Unifying Forces
- Understanding Reality Synchronization Through Crisis

16.2 Disasters as Dimensional Gateways

- The Mechanism of Reality Alignment During Collective Crises
- Shared Interests and Priorities in the Face of Danger
- How the Multiverse Temporarily Collapses into a Unified Dimension

16.3 Traveling Between Realities: Collective Experiences as Gateways

- How Collective Experiences Facilitate Dimensional Shifts
- The Role of Shared Focus in Collapsing Multiple Realities into One

16.4 Gradual Return to Individual Realities

- The Post-Disaster Fragmentation of Shared Dimensions
- How Personal Beliefs and Perceptions Reassert Themselves Over Time

16.5 Empathy and Shared Experience as Bridges Between Dimensions

- The Role of Empathy in Creating Temporary Dimensional Convergence
- Why Shared Experience Creates a Sense of Oneness Beyond Individual Beliefs

16.6 Disasters as Catalysts for Dimensional Awareness

- How Experiencing New Realities During Crises Expands Consciousness
- The Long-Term Impact of Temporarily Occupying a Shared Dimension

Conclusion: Understanding the Multiverse of Thought

- The Implications of the Perceptual Multiverse Theory
- Flashpoint Issues as Gateways to New Realities
- The Future of Humanity in a Multiverse of Infinite Dimensions

Introduction: The Perceptual Multiverse Unveiled

Defining the Multiverse of Thought and Belief

The idea of the multiverse has long been a tantalizing concept in both science and philosophy. Traditionally, the term **multiverse** refers to the possibility of multiple, perhaps infinite, universes existing alongside our own, each with its own unique physical laws, histories, and realities. It's a powerful and transformative concept that stretches the boundaries of our understanding of existence. But beyond the scope of **parallel physical realities**, there's another, equally profound interpretation of the multiverse—one that deals with the mind, consciousness, and belief.

This is what I call the **Perceptual Multiverse**: a multiverse not only of different physical realities but of different **mental dimensions**, each shaped by our beliefs, perceptions, and experiences. In the Perceptual Multiverse, **thought and belief** act as the fundamental forces that give shape to each dimension. The multiverse exists not just as a series of parallel physical worlds, but as a series of **parallel thought worlds**, each occupied by those whose perceptions align with that dimension's reality.

Imagine for a moment that every individual lives in their own **thought dimension**, where the nature of reality is determined by what they believe to be true. This dimension is shared with others who hold similar beliefs, creating a **collective experience** of reality that feels undeniably real to its inhabitants. Yet, in the Perceptual Multiverse, each of these dimensions exists alongside others—dimensions where vastly different beliefs and experiences create entirely separate realities.

The concept of the Perceptual Multiverse helps explain the **flashpoint issues** we see in the world today. These are moments where different **thought dimensions** intersect, leading to conflict, misunderstanding, and debate. Whether it's disagreements about the nature of climate change, political ideologies, or even fundamental issues of identity and existence, what we are truly witnessing are the collisions of different **mental dimensions**—parallel realities shaped by deeply held beliefs.

This theory offers a new lens through which to view the world. It suggests that the **conflicts** we experience are not merely the result of differing opinions but are the manifestations of people **living in separate realities**, each as real and valid to them as our own is to us. The Perceptual Multiverse proposes that **belief** has the power to **create dimensions**, and by understanding this, we can begin to see how people can inhabit different realities even while physically sharing the same world.

The Nature of Reality: Objective vs. Subjective Worlds

To understand the Perceptual Multiverse fully, we need to explore the fundamental question: **What is reality?** Traditionally, reality is understood in two primary ways: **objective reality** and **subjective reality**.

Objective Reality:

Objective reality refers to the world as it exists **independently of human thought or perception**. This is the realm of measurable facts, scientific laws, and physical phenomena that exist whether or not we observe them. In this view, reality is something **fixed** and **unchanging**, governed by the laws of physics and operating on principles that are universal. The Earth orbits the sun, water boils at 100 degrees Celsius, and gravity pulls objects toward the center of the Earth—these are facts that exist in the realm of objective reality, verified by empirical evidence and observation.

From the perspective of objective reality, there is a **single truth** that applies to all, regardless of belief or perception. Science and reason are the tools we use to uncover this truth, peeling back the layers of ignorance or misunderstanding to reveal the objective world beneath.

Subjective Reality:

Subjective reality, on the other hand, is the world as it is **experienced by individuals**. It is shaped by our thoughts, emotions, beliefs, and perceptions. Each of us navigates the world through the lens of our personal experiences, and as such, our subjective reality can be very different from the objective reality—or from the reality experienced by someone else.

In subjective reality, **truth** is not absolute. It is shaped by perspective. What is true for one person may not be true for another, and this is where the **Perceptual Multiverse** takes root. In this framework, **subjective truth** forms the foundation of entire dimensions of thought and experience. People who share similar beliefs inhabit the same **subjective reality**, creating a collective dimension that, for them, feels as real as the objective world.

For example, consider two people looking at a controversial political issue. One person might see the issue through the lens of **justice and equality**, while the other sees it as a matter of **personal freedom and responsibility**. Each person is inhabiting a different **subjective reality**, shaped by their beliefs, values, and experiences. To each of them, their perception of the issue is the truth, and they cannot easily understand how the other person could see things differently.

The clash between these **subjective realities** is what creates the **flashpoints** we see in society today. But these clashes are not just disagreements—they are the meeting points of different **dimensions** within the Perceptual Multiverse. In this multiverse, subjective realities are not simply different opinions or perspectives—they are entire **parallel worlds** that coexist alongside our own, shaped by the beliefs and perceptions of those who inhabit them.

The Role of Perception in Shaping Dimensions

Perception is the key to understanding how these dimensions come into being. It is through perception that we experience the world, interpret it, and ultimately create our version of reality. But perception is not a passive process—it is an active **shaper of experience**. How we perceive the world not only influences how we interact with it but also determines the **dimension** we inhabit.

In the Perceptual Multiverse, perception acts as the **gateway** to different dimensions. What we perceive as true defines the reality we experience, and those perceptions are often influenced by a combination of personal experience, cultural background, and shared belief systems.

For example, imagine a person who is deeply invested in the belief that **climate change is a hoax**. Every piece of information they encounter—whether it supports or refutes their belief—will be filtered through the lens of their perception. Information that aligns with their belief will be accepted as fact, while information that contradicts it will be dismissed or explained away. In this way, their perception creates a **dimension** where climate change is not real. They inhabit a reality where the overwhelming scientific consensus on climate change is seen as part of a grand conspiracy. This dimension is **real** to them, just as real as the dimension where climate change is accepted as an undeniable fact by those who perceive the world differently.

In this way, perception shapes the very fabric of the Perceptual Multiverse. It determines not only what we believe to be true but also the **reality we experience**. When enough people share the same perception, they create a **shared dimension**, a collective subjective reality where their beliefs are the foundation of that dimension's existence. This is why entire communities, cultures, or social movements can seem to inhabit **different worlds**, even though they physically share the same space.

Perception as a Gateway to Other Dimensions

But perception does more than shape the dimensions we currently inhabit—it also allows us to access **new dimensions**. By changing the way we perceive the world, we can **shift** from one dimension to another, exploring different realities that were previously beyond our reach.

For example, through **meditation**, individuals can alter their state of consciousness and access higher spiritual dimensions. In these altered states, they might experience profound insights, connections with spiritual entities, or even entire realms of existence that feel more real than the physical world. These experiences suggest that perception can act as a **portal** to dimensions that exist beyond our normal waking consciousness.

Similarly, **vibrational practices**—such as sound healing, energy work, or certain forms of yoga—can help individuals **raise their vibrational frequency**, allowing them to access dimensions of thought and experience that vibrate at a higher frequency than the physical world. In these dimensions, time and space may behave differently, and individuals can experience realities that are more fluid, interconnected, and expansive.

In the Perceptual Multiverse, **dimensions of thought and belief** are as real as any physical dimension. They exist alongside our objective reality, but they are accessible only through perception. The more we expand our perception, the more dimensions we can explore, and the more we understand the vast multiverse that lies beyond our everyday experience.

The Role of Belief in Shaping Reality

In the Perceptual Multiverse, **belief** is the driving force behind the creation of dimensions. It is through belief that we **anchor** ourselves in a particular reality. What we believe to be true shapes the **laws** of the dimension we inhabit, just as the laws of physics shape the behavior of objects in the physical world.

For example, a person who believes that success is determined by hard work will experience a dimension where effort leads to reward, and they will encounter opportunities that reinforce this belief. On the other hand, a person who believes that success is determined by luck or circumstance may experience a dimension where their efforts seem futile, and success appears to be out of their control.

Belief also determines how we **interact** with others. People who share the same beliefs will naturally gravitate toward one another, creating **collective dimensions** where their shared reality is reinforced. This is why communities form around shared ideologies, religions, or worldviews. These communities are **dimensional hubs** where belief acts as the foundation of the collective experience.

When people with **different beliefs** interact, it can feel like they are speaking different languages or living in different worlds—and in a very real sense, they are. They are each rooted in their own **belief dimension**, and unless they are willing to change their beliefs, they will remain anchored in that reality.

Chapter 1: Foundations of the Multiverse

1.1 Scientific Roots of the Multiverse Theory

The concept of the **multiverse**—the idea that multiple universes could exist simultaneously—has long fascinated both scientists and philosophers alike. While traditionally, cosmology and quantum mechanics have approached the multiverse as a collection of parallel physical worlds, this chapter explores how those same scientific principles can be extended to **perceptual realities**—a mental multiverse shaped by the thoughts, beliefs, and perceptions of individuals.

The Big Bang and Inflation Theory: The Birth of the Multiverse

The idea of a **multiverse** emerged as a solution to some of the puzzles left behind by the Big Bang theory. According to the Big Bang theory, the universe began from a hot, dense point around 13.8 billion years ago and has been expanding ever since. While this model successfully explains many observations—such as the cosmic microwave background radiation and the distribution of galaxies—it also presents certain mysteries. For instance, why does the universe appear so **uniform** on large scales? Why do regions of the universe that should never have been in contact with each other have such similar properties?

In the 1980s, physicist **Alan Guth** proposed a solution known as **cosmic inflation**. According to inflation theory, in the first fraction of a second after the Big Bang, the universe underwent a period of rapid, exponential expansion. This solved the problem of uniformity by suggesting that everything we see in the universe today came from a small, homogeneous region that inflated to cosmic proportions.

However, cosmic inflation also gave rise to the idea of a **multiverse**. If inflation happened once, why not assume that it could happen again in other regions of space? This leads to the idea of **eternal inflation**, where new universes are constantly being created in an ever-expanding sea of space. Each of these universes could have different physical laws, constants, and properties, making them **separate realities** from our own.

The Many-Worlds Interpretation of Quantum Mechanics

While cosmic inflation explains how multiple universes might arise from the structure of space itself, another branch of physics—**quantum mechanics**—suggests that multiple realities could arise from the very nature of matter and energy at the smallest scales. The **many-worlds interpretation** of quantum mechanics, first proposed by physicist **Hugh Everett III** in 1957, posits that every time a quantum event occurs, the universe **splits** into different branches, with each possible outcome happening in a separate, parallel universe.

At the heart of quantum mechanics is the idea that particles, like electrons, can exist in multiple states simultaneously—what physicists call **superposition**. For example, an electron can exist in multiple locations or spin in multiple directions at the same time. However, when we measure the particle, we always observe it in a specific state. According to the many-worlds interpretation, the act of measurement doesn't collapse the particle's superposition; instead, it causes the universe to **split** into multiple branches, one for each possible outcome of the measurement.

This creates an **infinite number of parallel universes**, each representing a different version of reality. In one universe, the electron is observed to be spinning up, while in another, it is spinning down. Every decision, every quantum event, creates a branching of the universe, leading to a **multiverse** of possibilities where every potential outcome plays out in a different reality.

The Multiverse and String Theory

Another major player in multiverse theory is **string theory**—an ambitious attempt to unify all of the fundamental forces of nature, including gravity, into a single framework. In string theory, the fundamental building blocks of the universe are not point-like particles but tiny, vibrating **strings**. The different vibrational modes of these strings give rise to the different particles we observe, from electrons to photons to quarks.

One of the most intriguing predictions of string theory is the existence of **extra dimensions**—beyond the familiar three dimensions of space and one dimension of time. These extra dimensions are "curled up" so small that they are imperceptible to us, but they are thought to be essential to the fundamental nature of the universe. However, string theory also opens the door to the existence of **multiple universes**, each with a different configuration of these extra dimensions. In one universe, gravity might be much stronger than in ours, while in another, electromagnetic forces might behave differently.

The idea of the multiverse has found fertile ground in both **cosmology** and **quantum mechanics**. While these theories began as ways to explain the physical nature of the universe, they have since opened up the possibility that there may be multiple versions of reality, each as real as our own, but governed by different rules.

1.2 From Quantum Physics to Perceptual Realities

The leap from the **multiverse of physical worlds** to a **multiverse of perceptual realities** might seem large at first glance, but the same principles that govern the behavior of particles at the quantum level can also be applied to human consciousness and perception.

Perception as a Quantum Event

In quantum mechanics, the act of **observation** plays a crucial role in determining the outcome of an experiment. Before an observation is made, particles exist in a state of superposition, where all possible outcomes exist simultaneously. It is only when an observer interacts with the system that one particular outcome is realized.

This raises a profound question: **What if human perception works the same way?** What if reality exists in a state of **superposition** until we observe it, and the act of observation itself determines which reality we experience? In other words, what if **perception** acts as the **mechanism** by which we select from the various possibilities offered by the multiverse?

This is the foundation of the **Perceptual Multiverse**—the idea that the reality we experience is shaped by our **perception**, and that each of us inhabits a different dimension of reality based on how we perceive the world. Just as particles exist in multiple states until they are observed, the **Perceptual Multiverse** posits that multiple realities exist simultaneously, and it is our **beliefs** and **perceptions** that determine which reality we experience.

In this framework, **thought** and **belief** are the quantum events that cause the "collapse" of reality into a particular form. When we believe something to be true, we are effectively selecting from the various possibilities offered by the multiverse and creating a reality where that belief is the governing principle. This is why people with different beliefs can seem to live in **entirely different worlds**, even though they share the same physical space. They are each **collapsing** the multiverse into a different version of reality based on their perceptions.

The Observer as a Creator of Reality

One of the most intriguing aspects of quantum mechanics is the role of the **observer**. In the classic **double-slit experiment**, particles behave differently depending on whether or not they are being observed. When the particles are not observed, they behave like waves, spreading out and passing through both slits simultaneously. But when an observer measures the particles, they behave like particles, passing through only one slit at a time.

This experiment suggests that the **act of observation** changes the behavior of particles, leading to the question: **What role does human perception play in creating reality?** If particles behave differently when they are observed, could the same be true of **larger systems**, such as human experiences, social structures, and even entire worlds?

In the Perceptual Multiverse, the **observer** is not just a passive witness to reality but an active **creator** of it. By perceiving the world in a particular way, the observer is effectively choosing from the multiple possibilities offered by the multiverse and creating a version of reality that aligns with their beliefs and perceptions. This means that each person is **living in their own reality**, shaped by their thoughts, beliefs, and experiences.

For example, someone who believes in a **just and fair world** will experience a reality where events and circumstances seem to align with that belief. In this dimension, people are rewarded for their hard work, justice is served, and fairness prevails. Meanwhile, someone who believes that the world is **unfair and chaotic** will experience a different dimension, where injustice and randomness seem to govern the world. Both realities exist simultaneously in the multiverse, but the **observer's perception** determines which one they experience.

This is why **flashpoint issues** in society can seem so intractable. People who disagree on fundamental issues are not just holding different opinions—they are living in **different realities**. Their perceptions have shaped their version of reality to such an extent that they can no longer see the world from the perspective of someone who inhabits a different dimension.

1.3 The Observer Effect and Reality Creation

The Observer Effect in Physics

The **observer effect** in quantum mechanics refers to the idea that the mere act of observing a system can change its behavior. This is most famously demonstrated in the double-slit experiment, where the presence of an observer causes particles to behave differently than they would if they were not being observed.

This effect is not limited to the quantum world. It has profound implications for the nature of **reality** itself. If observing a system can change its behavior, what does that say about the role of **consciousness** in shaping the world? Could it be that our thoughts and perceptions are not just passive reflections of reality, but active forces that shape the very fabric of existence?

The Perceptual Observer Effect: Shaping Thought Dimensions

In the **Perceptual Multiverse**, the observer effect extends beyond the quantum level to include the creation of entire **thought dimensions**. Just as the act of observation in quantum mechanics determines the behavior of particles, the act of **perception** in the Perceptual Multiverse determines the nature of the dimension we inhabit.

When we perceive the world in a particular way, we are not merely **observing** reality—we are **creating** it. Our perceptions act as the **catalyst** that brings a particular dimension into existence. In this sense, perception is not just a passive process but an active **force** that shapes the multiverse.

For example, someone who perceives the world as a place of opportunity and abundance will experience a dimension where opportunities seem to abound, and success comes easily. In contrast, someone who perceives the world as a place of scarcity and struggle will experience a dimension where resources are limited, and success is hard to come by. Both dimensions exist in the multiverse, but the observer's perception determines which one they inhabit.

Collapsing the Multiverse: How Perception Selects Reality

In the Perceptual Multiverse, perception acts as the mechanism by which we **collapse** the multiverse into a particular reality. Just as a quantum system exists in a state of superposition until it is observed, the multiverse exists as a collection of **possible realities** until our perception selects one of them.

This idea can be extended to explain why different people experience the world in such **different ways**. When we perceive the world through a particular lens—whether it be one of optimism, pessimism, justice, or chaos—we are effectively **collapsing** the multiverse into a dimension that aligns with that perception.

This process of collapsing the multiverse into different dimensions explains why **flashpoint issues** arise in society. People with different beliefs and perceptions are not just experiencing different aspects of the same reality—they are experiencing **different realities** altogether. Their perceptions have shaped the world they inhabit, and because these worlds are rooted in different dimensions, they often seem incompatible.

Bridging Science and Thought: Perception as Reality

By exploring the roots of **multiverse theory** in physics and applying the same principles to human **perception**, we can begin to see how the **Perceptual Multiverse** functions. Just as quantum mechanics suggests that the act of observation shapes the outcome of an event, the Perceptual Multiverse suggests that **belief** and **perception** shape the dimensions we inhabit.

In the chapters that follow, we will explore how these perceptual dimensions manifest in the world through **flashpoint issues**, and how understanding the Perceptual Multiverse can help us navigate the complexities of **parallel realities**.

Chapter 2: Dimensions Beyond Perception

2.1 What Are Extra Dimensions?

The Basic Concept of Dimensions

To understand the concept of **extra dimensions**, we first need to define what a **dimension** is. In physics, dimensions refer to the different aspects of space and time that we can measure. We live in a world with three spatial dimensions: length, width, and height, which we can observe and navigate daily. These three dimensions, combined with the dimension of time, create what physicists call **spacetime**—the fabric of the universe as we know it.

However, the idea of **extra dimensions** suggests that there could be more than these familiar four. These additional dimensions might be hidden from our perception, existing in ways that we cannot easily observe or experience. In modern physics, extra dimensions are a key feature of theories like **string theory** and the **multiverse theory**, both of which propose that our universe is only one of many, and that the laws of physics we experience might be a subset of a much larger reality that includes higher dimensions.

Extra Dimensions in Science: String Theory and Beyond

One of the most popular frameworks that includes extra dimensions is **string theory**, which posits that the fundamental building blocks of the universe are not particles like electrons or quarks but instead tiny, vibrating strings of energy. The way these strings vibrate determines the properties of particles and forces. According to string theory, for this framework to work, there must be more than four dimensions. In fact, string theory requires up to **ten or eleven dimensions** for the mathematics to be consistent.

But why don't we perceive these extra dimensions? According to string theorists, these additional dimensions are **curled up** so small that they are beyond the reach of our current observational capabilities. Imagine a two-dimensional sheet of paper. If you roll the paper into a tight tube, it may appear one-dimensional from a distance, but closer inspection reveals that it still has two dimensions. Similarly, these extra dimensions are thought to be so tightly compacted that we don't notice them in our daily lives.

Why is this important? If these extra dimensions exist, they might have profound effects on the nature of the universe. They could explain phenomena we haven't yet understood, from the behavior of fundamental forces to the ultimate fate of the cosmos. In the context of the **Perceptual Multiverse Theory**, these extra dimensions could also represent **alternate realities** or **planes of existence** that are accessible not through traditional physics, but through perception, thought, or altered states of consciousness.

Dimensions in Quantum Mechanics: Beyond the Visible

Quantum mechanics adds another layer of complexity to the idea of dimensions. At the quantum level, particles behave in ways that defy our understanding of the classical world. They can exist in multiple states at once (superposition), can become "entangled" with other particles across vast distances, and seem to only solidify into a definite state when they are **observed**. This leads to the idea that **multiple possibilities**—multiple realities—might exist simultaneously, even though we only perceive one outcome.

When applied to the Perceptual Multiverse, quantum mechanics suggests that **different dimensions** of thought and experience might exist **simultaneously**—just like quantum superposition states—but we only perceive the reality that aligns with our current state of consciousness or belief. By changing how we perceive the world, we may be able to access **hidden dimensions** of thought or existence that have always been present but were previously beyond our awareness.

In quantum mechanics, there's also the concept of **probability waves**—that is, particles don't exist in a single point but rather as a cloud of probabilities, with each possibility existing simultaneously until an observation "collapses" the wave function into a specific outcome. This can be extended to how we understand **thought dimensions**: multiple potential realities exist, but it's our **perception**—our observation—that determines which reality we experience.

Cultural and Philosophical Interpretations of Extra Dimensions

While modern physics provides a mathematical and theoretical basis for the existence of extra dimensions, cultures around the world have long entertained the idea that there are realities beyond what we can perceive with our senses. Many religious and spiritual traditions describe otherworldly realms—**heaven**, the **astral plane**, the **spirit world**—that exist alongside or beyond our physical reality. These could be seen as **perceptual dimensions**, accessible not through scientific instruments but through altered states of consciousness, meditation, or spiritual practice.

In some Indigenous and Eastern traditions, for example, shamans, monks, or spiritual practitioners enter other dimensions through rituals, meditation, or natural substances that alter their perception. These altered states allow them to access dimensions where they can communicate with spirits, receive wisdom, or heal ailments. From a scientific perspective, these realms might be thought of as **mental or spiritual dimensions**, not unlike the extra dimensions described in string theory, but they are accessed through the **mind** rather than physical instruments.

This leads us to a key question: **What if these extra dimensions exist not only in the realm of physics but also in the realm of consciousness?** What if, by expanding our perception, we could access **invisible realities** that have always been present but were previously beyond our reach?

2.2 Invisible Realities and the Limits of Human Perception

The Biological Limitations of Perception

As humans, our perception of reality is constrained by our **biological senses**. Our eyes can only detect a narrow band of the electromagnetic spectrum, which we perceive as visible light. Our ears can only pick up a limited range of sound frequencies. Our sense of touch, taste, and smell are similarly limited in scope. In short, we are biologically designed to perceive only a small fraction of what might exist in the universe.

For example, **infrared light** exists, but we cannot see it without special equipment. **Ultrasound** exists, but we cannot hear it. Similarly, it's entirely possible that there are **other dimensions** or **realities** that exist just beyond our sensory capacities. Just as we need instruments to detect infrared light or radio waves, we may need new tools—perhaps mental or spiritual tools—to perceive these extra dimensions.

The brain itself is an incredibly complex organ, but it is not infallible. It takes in massive amounts of sensory data from the environment and filters out what it deems unnecessary. This filtering process is crucial for survival, but it also means that much of what we could potentially perceive is ignored or discarded by the brain before it reaches our conscious awareness.

In this way, human perception acts as a **narrow window** through which we experience reality. What lies beyond this window could be vast, complex, and profoundly different from the reality we know. These **invisible realities** are not inherently inaccessible; they are simply beyond the reach of our biological senses and the standard frameworks we use to understand the world.

The Role of Consciousness in Perceiving Hidden Realities

While our biological senses may limit our perception of reality, **consciousness** offers a different avenue for exploration. In various philosophical, spiritual, and scientific traditions, consciousness is seen as more than just a passive observer of reality—it is an **active creator** and **shaper** of reality.

In the **Perceptual Multiverse Theory**, consciousness is the key to accessing **hidden dimensions**. By altering our state of consciousness—whether through meditation, deep introspection, or certain technologies—we can expand our perception and begin to experience realities that were previously invisible to us.

Altered states of consciousness have been used throughout history as a means of accessing different dimensions of reality. For example, in **Tibetan Buddhism**, advanced practitioners use meditation to access higher states of consciousness, where they can perceive subtle realms beyond the physical world. Similarly, **shamans** in Indigenous cultures use trance states to journey to other realms, where they interact with spirits and gain insights that are inaccessible in ordinary waking life.

In modern psychology, the concept of **neuroplasticity**—the brain's ability to change and adapt—suggests that even our everyday perception of reality is not fixed. Through focused mental effort, we can **train the brain** to perceive the world in new ways, opening the door to experiences and realities that were once beyond our comprehension.

The limits of human perception are not hard boundaries. They are **thresholds** that can be expanded through training, practice, and the deliberate cultivation of **altered states of consciousness**. By learning to transcend our biological limitations, we can begin to explore the **hidden dimensions** that exist beyond our current perception.

Tools and Technologies for Expanding Perception

Throughout history, humans have developed tools to extend the range of their perception, allowing them to explore **invisible realities** that would otherwise remain hidden. Telescopes and microscopes have revealed worlds at scales we could never see with the naked eye, while instruments like spectrometers and particle detectors have opened up entirely new realms of scientific inquiry.

In recent years, **neurotechnology** has begun to explore the boundaries of **consciousness itself**, providing new ways to expand and alter human perception. Technologies such as **neurofeedback**, **virtual reality**, and **brainwave entrainment** are being used to explore altered states of consciousness, allowing individuals to tap into deeper levels of awareness and potentially access **hidden dimensions** of thought and experience.

For example, **virtual reality (VR)** creates entirely new environments that can trick the brain into believing it is experiencing a different reality. While current VR technology primarily simulates physical environments, future developments may allow us to create **virtual dimensions** that are based on the **subjective experience** of consciousness, allowing us to explore new realms of thought and emotion.

Another emerging technology is **brainwave entrainment**, which uses specific frequencies of sound and light to synchronize the brain's electrical activity with a desired frequency. By tuning the brain to certain frequencies, individuals can induce **altered states of consciousness** that may allow them to perceive realities beyond their ordinary waking experience.

While these technologies are still in their infancy, they represent the beginning of a new frontier in human exploration—one that goes beyond the physical world and into the **dimensions of thought** and **perception** that make up the **Perceptual Multiverse**.

2.3 Unlocking the Potential of Future Dimensions

The Expanding Horizon of Human Perception

As we continue to explore the boundaries of human perception, the potential for discovering **new dimensions of reality** grows ever larger. Just as the development of telescopes and microscopes expanded our understanding of the physical universe, the ongoing exploration of consciousness and perception holds the promise of revealing **new realms** of existence that are currently hidden from view.

These future dimensions may not be physical in the traditional sense. They may exist as **mental realms**, **vibrational frequencies**, or **spiritual planes** that are accessible only through altered states of consciousness or advanced technologies. Yet they are no less real than the physical universe we inhabit. They are simply **different aspects** of the larger multiverse, waiting to be discovered and explored.

Dimensions of Vibrational Frequencies

One of the most intriguing possibilities for future dimensions lies in the concept of **vibrational frequencies**. In many spiritual traditions, it is believed that all matter and energy vibrates at different frequencies, and that by raising or lowering one's vibrational frequency, it is possible to access different planes of existence.

In the **Perceptual Multiverse Theory**, these vibrational dimensions represent different levels of reality that exist in parallel with our own. Each vibrational dimension operates at a specific frequency, and by tuning into that frequency, we can experience the **reality** that exists at that level.

For example, the physical world we experience is thought to vibrate at a relatively low frequency, while higher dimensions—such as the **astral plane** or **etheric realm**—vibrate at higher frequencies. These higher dimensions are often described as more fluid, interconnected, and responsive to thought, where the boundaries between **mind** and **matter** are blurred.

By learning to **raise our vibrational frequency**, we may be able to access these higher dimensions and experience realities that are more aligned with **thought** and **consciousness**. Practices such as meditation, sound therapy, and energy work are often used to help individuals attune to higher frequencies and open the door to these hidden dimensions.

Mental Dimensions: The Realm of Thought and Belief

Another frontier in the exploration of future dimensions is the **realm of thought and belief**. In the **Perceptual Multiverse**, thought and belief are not passive reflections of reality—they are **active forces** that shape and create the dimensions we inhabit.

As we begin to understand the power of thought in shaping reality, we can unlock the potential to create and explore **mental dimensions** that are entirely based on our beliefs and perceptions. These dimensions are not bound by the physical laws of our universe but are instead governed by the **laws of thought**.

For example, a person who believes in the power of positive thinking may experience a dimension where their thoughts manifest as opportunities, success, and abundance. In contrast, someone who believes in the inevitability of struggle may experience a dimension where obstacles and challenges are constant.

These **mental dimensions** are not just theoretical—they are **lived realities** for those who inhabit them. By understanding the role of thought in shaping these dimensions, we can begin to consciously choose the dimensions we wish to inhabit, and create new realities based on our evolving beliefs and perceptions.

Chapter 3: Flashpoint Issues as Dimensional Evidence

3.1 Defining Flashpoint Issues: Clashes of Parallel Realities

Flashpoint issues refer to critical points of conflict or tension in society where groups of people become deeply divided, often over fundamental disagreements about what is true, real, or morally right. These issues are typically high-stakes, emotionally charged, and resistant to easy resolution. Examples include debates over climate change, gender identity, vaccination, abortion rights, and racial inequality. Each of these issues represents a point where **parallel realities**—or **thought dimensions**—collide, making it clear that different groups are experiencing entirely different versions of reality.

In the framework of the **Perceptual Multiverse Theory**, flashpoint issues serve as **dimensional clashes** where people who occupy different **thought dimensions** come into direct conflict. These conflicts arise not because one side is necessarily "right" or "wrong" in an absolute sense, but because each group is operating from a distinct **dimensional perspective**, shaped by their beliefs, experiences, and perceptions.

Flashpoints as Dimensional Collisions

Consider a flashpoint issue like **climate change**. One group of people lives in a reality where the evidence for climate change is overwhelming and undeniable, with immediate and severe consequences if action is not taken. This group occupies a dimension where the **scientific consensus** is the dominant narrative, and their reality is shaped by data on rising temperatures, melting ice caps, and extreme weather events.

Meanwhile, another group occupies a parallel dimension where climate change is either not occurring, or its severity is vastly overstated. In this dimension, skepticism about the scientific narrative is the norm, and the focus is on economic concerns or the belief that environmental policies are a form of control or manipulation. To this group, the data on climate change is interpreted through a lens of mistrust, and their version of reality downplays or outright denies the need for urgent action.

These two dimensions are **fundamentally incompatible**. Both groups believe they are experiencing the "real" world, but in fact, they are each occupying separate thought dimensions, where the same information is perceived and processed differently. The **flashpoint** occurs when these two groups come into contact, often through media, politics, or social activism, leading to an intense clash of realities.

Flashpoint issues are not limited to large-scale societal debates. They can occur in smaller, more personal contexts as well, such as within families, workplaces, or communities. Whenever people with different **perceptual realities** encounter one another, there is the potential for a flashpoint, as their **thought dimensions** intersect in ways that create conflict.

Characteristics of Flashpoint Issues

Flashpoint issues tend to have several defining characteristics:

1. **Polarization**: Flashpoint issues are deeply divisive, often creating stark divisions between groups. These divisions are typically driven by differences in belief systems, values, and perceptions, with each group viewing the other as fundamentally misguided or wrong.
2. **Emotional Intensity**: Flashpoints are charged with emotional energy, as the stakes are often perceived to be incredibly high. Whether the issue is framed as a matter of justice, survival, freedom, or morality, people on both sides feel strongly invested in the outcome, which only deepens the conflict.
3. **Perceived Irreconcilability**: Flashpoint issues often feel intractable because they represent more than simple disagreements over facts or policy. Instead, they reflect **fundamental differences in reality**—each group is operating from a different **thought dimension**, and as a result, it can be nearly impossible to find common ground.
4. **Symbolic Significance**: Flashpoints are not just about the specific issue at hand—they often serve as symbols for larger conflicts. For example, debates over **vaccination** may represent deeper conflicts about trust in science, individual autonomy, and government authority.

By recognizing flashpoint issues as **dimensional clashes**, we can begin to understand why these conflicts are so resistant to resolution. It is not just a matter of convincing one side of the other's perspective, but rather a question of how to navigate the intersection of **parallel realities**.

3.2 Why Flashpoints Are Proof of Different Thought Dimensions

Flashpoint issues offer clear **proof** of the existence of different **thought dimensions** within society. These dimensions are not merely ideological or opinion-based; they are **full-fledged realities** experienced by those who inhabit them. The intensity of flashpoint issues arises because these dimensions are fundamentally incompatible, yet they exist side by side in the same physical world.

Thought Dimensions: Shaping Perception and Reality

In the **Perceptual Multiverse, thought dimensions** are mental and emotional realities shaped by a person's beliefs, experiences, and cognitive frameworks. These dimensions govern how individuals interpret information, what they prioritize as important, and how they make sense of the world around them.

For example, a person who inhabits a dimension where **personal freedom** is the highest value will perceive issues like **mask mandates** or **vaccination requirements** through the lens of individual autonomy. In this reality, any restriction on personal choice is seen as an infringement on fundamental rights, and the government is viewed as overstepping its bounds. Conversely, a person who inhabits a dimension where **public health** and **community well-being** are the highest priorities will see these same mandates as necessary protections against a collective threat. In this dimension, the focus is on protecting vulnerable populations, and government intervention is seen as a necessary measure to ensure public safety.

Both of these individuals are experiencing **different dimensions of reality**, shaped by their respective values and beliefs. The flashpoint arises when these realities come into conflict, and each person perceives the other's reality as fundamentally wrong or dangerous. From the outside, it may appear that these individuals are simply disagreeing over policy, but in reality, they are clashing over **competing dimensions** of thought and experience.

Cognitive Dissonance as Dimensional Evidence

Cognitive dissonance—the mental discomfort experienced when holding conflicting beliefs or encountering information that contradicts one's worldview—provides further evidence of the existence of **thought dimensions**. When a person is confronted with a flashpoint issue that challenges their existing reality, they often experience cognitive dissonance as their mind struggles to reconcile conflicting dimensions.

For example, a person who strongly believes in the safety and efficacy of vaccines may experience cognitive dissonance when interacting with someone who believes that vaccines are harmful or part of a conspiracy. The clash between these **thought dimensions** creates a mental and emotional strain, as each person's perception of reality is called into question.

This dissonance is a signal that the individual is encountering a **dimensional conflict**—they are being forced to confront the existence of a parallel reality that operates on different rules and assumptions. The more deeply entrenched a person is in their thought dimension, the more intense the cognitive dissonance when they encounter conflicting realities.

In this way, **flashpoint issues** not only highlight the existence of different thought dimensions but also reveal the difficulty of **navigating** between these dimensions. Cognitive dissonance acts as a kind of **dimensional tension**, signaling that the person is moving between conflicting realities, even if only briefly.

The Limits of Objective Truth in Thought Dimensions

One of the challenges in addressing flashpoint issues is that they often defy resolution through **objective truth**. In many cases, individuals from different thought dimensions are not swayed by facts, data, or evidence that contradict their beliefs. This is because, within their dimension, their perception of reality is **self-reinforcing**—they selectively filter information to fit their existing worldview, and any contradictory evidence is either dismissed or reinterpreted to align with their beliefs.

For example, in the debate over **climate change**, scientific data overwhelmingly supports the conclusion that human activity is driving global warming. However, individuals who inhabit a dimension where climate change is not perceived as a serious threat will often reject this data, viewing it as flawed, manipulated, or part of a broader political agenda. In their dimension, the "truth" about climate change is fundamentally different from the truth perceived by those who accept the scientific consensus.

This selective filtering of information is not necessarily a conscious process—it is a natural consequence of living in a **self-contained dimension of thought**. Each dimension operates with its own internal logic, and within that logic, the individual's beliefs are consistent and rational. The difficulty arises when people from different dimensions attempt to communicate, as they are effectively speaking from **different realities** with **different truths**.

Flashpoint issues demonstrate the **limitations of objective truth** in resolving conflicts between thought dimensions. While facts and evidence are important, they are often insufficient to bridge the gap between competing realities. This is because the conflict is not just about the facts themselves, but about the **dimensional framework** through which those facts are interpreted.

3.3 How Dimensions Overlap and Intersect in Society

While thought dimensions are often self-contained, they do not exist in isolation. In the **Perceptual Multiverse**, dimensions of thought frequently **overlap and intersect**, creating areas of conflict but also opportunities for understanding and cooperation. These intersections are the spaces where people with different realities come into contact, and it is in these moments that flashpoint issues become most visible.

Dimensional Overlap in Public Spaces

One of the most common places where **dimensional overlap** occurs is in **public spaces**—whether physical spaces like workplaces, schools, or public forums, or digital spaces like social media platforms. In these environments, people from different thought dimensions are brought together, often without realizing that they are operating from entirely different realities.

For example, in a workplace setting, employees might come from different **political** or **cultural dimensions**, with vastly different views on issues like diversity, environmental sustainability, or corporate responsibility. These differences may not be immediately apparent, but when a flashpoint issue arises—such as a company policy on environmental practices or workplace diversity—the underlying **dimensional conflicts** are brought to the surface.

In these moments of dimensional overlap, people are forced to confront the fact that others perceive reality in fundamentally different ways. This can lead to conflict, but it can also create opportunities for **dimensional exchange**, where individuals begin to understand and explore the realities of others. While this process is often uncomfortable, it is also one of the few ways that people can begin to **navigate** between different dimensions and find common ground.

Social Media and the Amplification of Dimensional Conflict

Social media platforms have become a key space for **dimensional intersection**, but they often amplify rather than resolve **flashpoint conflicts**. On platforms like Twitter, Facebook, and Instagram, people from different thought dimensions come into contact in ways that are often **confrontational** and **polarizing**.

The **algorithmic nature** of social media tends to reinforce existing thought dimensions by curating content that aligns with the user's beliefs, creating **echo chambers** where individuals are exposed primarily to information that supports their existing reality. However, when content from other dimensions breaks through—such as a post or tweet from someone with a different worldview—the result is often a flashpoint conflict, as the user's reality is suddenly challenged by a competing dimension.

For example, a person who inhabits a dimension where **government regulation** is seen as a positive force for social good might encounter a post from someone who believes that government intervention is inherently harmful. The clash between these dimensions is intensified by the **public nature** of social media, where disagreements are often broadcast to large audiences and become part of a broader narrative of polarization.

While social media can exacerbate flashpoint conflicts, it also provides a unique opportunity for **dimensional awareness**. By exposing individuals to different realities, social media platforms create moments of **dimensional exchange**, where people can begin to understand the existence of other thought dimensions, even if they do not immediately agree with them.

Dimensional Intersections in Times of Crisis

As discussed in **Chapter 16**, times of crisis—such as natural disasters, economic recessions, or pandemics—can create moments of **dimensional convergence**, where people who usually inhabit separate realities are temporarily aligned in their focus and priorities. In these moments, thought dimensions overlap as individuals are forced to confront shared challenges and work together toward common goals.

For example, during the early months of the COVID-19 pandemic, people from different thought dimensions—whether focused on public health, personal freedom, or economic survival—were brought into a shared reality where the immediate concern was the health and safety of their communities. While these dimensions may have diverged as the pandemic wore on, the initial moment of crisis created a temporary alignment of realities, where people who normally disagreed found themselves united in their efforts to respond to a collective threat.

These moments of **dimensional intersection** highlight the potential for **collaboration** between thought dimensions, even in the face of deep-seated differences. They demonstrate that while flashpoint issues can be sources of intense conflict, they also offer opportunities for **dimensional exchange**, where individuals can gain insight into the realities of others and find ways to navigate between competing perceptions of the world.

Conclusion: Navigating Flashpoint Issues in the Perceptual Multiverse

Flashpoint issues serve as powerful evidence of the existence of **different thought dimensions** within society. These issues arise when people from **parallel realities** come into contact, creating moments of intense conflict and cognitive dissonance. The **Perceptual Multiverse Theory** offers a framework for understanding these conflicts not as simple disagreements but as clashes between **competing dimensions** of thought and belief.

By recognizing that flashpoint issues are manifestations of **dimensional intersections**, we can begin to approach these conflicts with greater empathy and understanding. Rather than trying to "win" the argument by imposing our reality on others, we can seek to **explore** and **navigate** the dimensions of those we disagree with, creating opportunities for **dimensional exchange** and **collaborative problem-solving**.

Ultimately, the **Perceptual Multiverse** reminds us that there are many ways to perceive and experience the world, and that each of these realities is valid within its own dimensional framework. By learning to navigate the **intersections** of thought dimensions, we can better understand the complexity of the human experience and work toward solutions that honor the diversity of realities that exist within our shared world.

Chapter 4: The Environmental Dimension

4.1 Climate Change: The Battle of Two Realities

The Scientific Reality of Climate Change

Climate change is one of the most pressing issues of the 21st century, and yet, it is also one of the most divisive. From a **scientific perspective**, climate change refers to the long-term alteration of temperature and typical weather patterns in a place. The overwhelming consensus among scientists is that human activities—primarily the burning of fossil fuels, deforestation, and industrial practices—are driving this change by increasing concentrations of greenhouse gases in the atmosphere.

This scientific understanding of climate change is supported by decades of data and research. The **Intergovernmental Panel on Climate Change (IPCC)**, for instance, has published numerous reports that document the rise in global temperatures, the melting of ice caps, the increase in sea levels, and the growing frequency of extreme weather events like hurricanes, wildfires, and droughts. The scientific reality of climate change is measurable and observable, and it suggests that if we do not take swift and meaningful action, we could face devastating consequences in the decades to come.

In this dimension, where climate change is seen as an **urgent crisis**, the focus is on finding solutions. This includes reducing carbon emissions, transitioning to renewable energy sources, and developing technologies that can mitigate the effects of climate change. Policy discussions in this dimension often revolve around **international cooperation** (as seen in agreements like the **Paris Climate Accord**), **corporate responsibility**, and the need for a fundamental shift in how societies operate to become more sustainable. In this reality, the future is precarious, and the choices made today will determine whether we avoid the worst effects of climate change or face irreversible damage to our planet.

The Skeptical Reality of Climate Change

In stark contrast to the scientific reality, there exists another dimension where **climate change skepticism** prevails. In this reality, people either **deny the existence of climate change** altogether or **minimize its significance**. The reasons for this skepticism are varied and complex, but they often include **political, economic, and cultural** factors that shape the way individuals and groups perceive environmental issues.

For many climate change skeptics, the idea that human activity could fundamentally alter the Earth's climate seems implausible. In this dimension, the Earth is seen as **too vast and too resilient** to be significantly impacted by human actions. Skeptics may point to natural climate cycles—such as the Ice Ages—as evidence that climate change is a natural phenomenon, not one caused by human behavior.

Furthermore, this dimension is often shaped by **economic concerns**. People who live in this reality may view efforts to combat climate change—such as regulating industries, transitioning to renewable energy, or implementing carbon taxes—as threats to their livelihoods. For example, individuals who work in industries like **coal mining, oil drilling**, or **manufacturing** may see environmental regulations as harmful to their jobs and communities. In this reality, the focus is on economic survival and personal freedom, rather than on the collective environmental impact of human activities.

In this dimension, skepticism about climate change is also often fueled by **political ideologies**. For example, in some conservative political frameworks, climate change is seen as a **left-wing agenda** aimed at expanding government control over the economy and restricting individual freedoms. As a result, climate change is viewed not as a scientific issue, but as a **political issue**, and efforts to address it are framed as threats to personal liberty and free enterprise.

The Clash Between Dimensions

The **clash between these two dimensions** is one of the defining flashpoints of our time. On one hand, you have a dimension where climate change is seen as an urgent, existential crisis requiring immediate action on a global scale. On the other hand, you have a dimension where climate change is either denied or downplayed, with a focus on economic growth and political freedom.

This clash plays out in public discourse, politics, and even personal relationships. People from the two dimensions often struggle to communicate with one another because they are not just disagreeing about policy—they are living in **different realities**. In one dimension, the evidence of climate change is overwhelming and inescapable, while in the other dimension, that same evidence is either ignored, dismissed, or reinterpreted to fit a different narrative.

The **Perceptual Multiverse Theory** helps explain why these realities are so difficult to reconcile. People in the climate change dimension are operating from a framework where **scientific evidence** and the need for collective action are paramount. Meanwhile, those in the skeptical dimension are operating from a framework that prioritizes **individual freedom, economic growth,** and a distrust of large-scale interventions. These dimensions are so divergent that it often seems impossible to find common ground, leading to the intense polarization we see today.

4.2 Experiencing the Environmental Crisis in Different Dimensions

Living in the Reality of Climate Crisis

For those who inhabit the **climate crisis dimension**, the evidence of climate change is not just a distant scientific concept—it is something they experience in their daily lives. These individuals are often acutely aware of the changes happening in the environment around them, whether it's **rising temperatures, changes in seasonal patterns,** or the increasing frequency of extreme weather events.

In this dimension, people are often deeply concerned about the future. They may feel a sense of urgency, even **existential dread**, as they contemplate the potential consequences of climate change. This can lead to a feeling of helplessness or frustration, especially when they see that not enough is being done to address the crisis.

Many who live in this dimension are activists, scientists, or individuals who have devoted their lives to promoting **environmental sustainability**. They are the ones pushing for **climate action** on a global scale, advocating for **policy changes**, and working to educate others about the dangers of continuing on our current path.

For example, the environmental movement known as **Extinction Rebellion** is rooted in this dimension. Members of this movement believe that drastic action is necessary to prevent a **climate catastrophe**, and they engage in acts of civil disobedience to raise awareness about the urgency of the issue. In this dimension, climate change is not a distant threat—it is an immediate, present-day crisis that requires bold, decisive action.

People in this dimension often make **lifestyle changes** in response to their understanding of climate change. This might include adopting a **plant-based diet**, reducing **carbon footprints** through minimal travel or energy use, or investing in **renewable energy** for their homes. For these individuals, the environmental crisis is a reality that shapes their daily decisions, from what they buy to how they live.

Living in the Skeptical Dimension

In contrast, those who inhabit the **skeptical dimension** of climate change experience the world very differently. While they may acknowledge that environmental issues exist, they do not see them as being as **immediate or severe** as those in the climate crisis dimension do. For them, the rhetoric around climate change can feel exaggerated or alarmist, and they often view environmentalism as an ideology rather than a science-based response to a pressing global issue.

In this dimension, people are more focused on the **economic consequences** of climate policies. They may worry about job losses in industries like fossil fuels, or they may believe that transitioning to renewable energy will lead to **higher costs** for consumers and businesses. For many in this dimension, the focus is on maintaining a **strong economy**, which they see as essential to their quality of life and personal freedom.

Moreover, in this dimension, people may question the motives of those who push for climate action. They might believe that **climate scientists** and activists are part of a broader agenda to restrict freedoms, increase government control, or push for a socialist economic system. As a result, they are resistant to calls for action, viewing them as part of a larger political or cultural war rather than a response to a genuine environmental threat.

In their daily lives, people in the skeptical dimension may not feel the same urgency or pressure to change their behavior in response to climate concerns. They might see **recycling** as a personal choice rather than a moral imperative, and they might be less likely to support policy initiatives that promote renewable energy or regulate carbon emissions. In this dimension, the focus is on **personal responsibility** and **economic growth**, rather than collective action to address a global problem.

The Emotional Experience of Different Dimensions

The **emotional landscape** of these two dimensions is vastly different. In the climate crisis dimension, people often feel a sense of **urgency, frustration,** and even **fear** about the future. They are deeply concerned about the potential for widespread ecological collapse, and they may experience anxiety or despair when they see that others do not share their concern.

In contrast, people in the skeptical dimension may feel a sense of **frustration** or **resentment** toward those who are pushing for climate action. They may see environmentalists as **out of touch** with the realities of economic survival, or they may feel that their personal freedoms are being threatened by what they perceive as **overreaching government policies**. For them, the focus is on protecting their way of life and resisting what they see as unnecessary or harmful changes.

These emotional experiences are a key part of why **climate change** has become such a polarizing issue. The clash of dimensions is not just about facts and data—it's about **values, emotions,** and deeply held beliefs about the world. This makes it incredibly difficult to bridge the divide between the two realities, as both sides are coming from fundamentally different emotional and cognitive frameworks.

4.3 The Future of Climate Dimensions: Collapse or Recovery?

The Potential for Dimensional Collapse

As we look to the future, the **Perceptual Multiverse Theory** suggests that the environmental dimension could either move toward **collapse** or **recovery**, depending on the actions we take in the coming years. One possible future is the dimension of **collapse**, where climate change continues unchecked, leading to widespread environmental degradation, social unrest, and economic instability.

In this dimension, we see the **worst-case scenarios** play out: **sea levels rise**, displacing millions of people from coastal cities; extreme weather events become more frequent and devastating, destroying homes, infrastructure, and ecosystems; and **food and water shortages** lead to conflict and competition for resources. This dimension is one where humanity fails to address the climate crisis, and as a result, we experience a cascade of ecological and societal breakdowns.

The **collapse dimension** is not just a theoretical possibility—it is a future that is already beginning to take shape. We are already seeing signs of this dimension in the form of **record-breaking heat waves**, devastating wildfires, and unprecedented levels of species extinction. If we continue on our current path, this dimension may become our lived reality, where the consequences of climate inaction are no longer hypothetical but painfully real.

The Dimension of Recovery

On the other hand, there is also the possibility of a **dimension of recovery**, where humanity rises to the challenge of climate change and takes bold action to **mitigate its effects**. In this dimension, we see a massive global effort to transition to **renewable energy**, reduce carbon emissions, and protect and restore ecosystems. Governments, corporations, and individuals work together to create a more **sustainable world**, where the focus is on long-term environmental health rather than short-term economic gain.

In the dimension of recovery, we can imagine a world where **green technology** flourishes, where new innovations in energy, transportation, and agriculture allow us to reduce our impact on the planet. In this dimension, international cooperation becomes the norm, and countries work together to develop solutions that benefit both people and the planet.

The recovery dimension is not without its challenges. It requires a fundamental shift in how we think about **growth, prosperity,** and the role of humans in the natural world. However, it also holds the promise of a brighter, more sustainable future, where we learn to live in harmony with the Earth rather than at odds with it.

Which Dimension Will We Choose?

The **future of climate dimensions** depends on the choices we make today. As we stand at the crossroads between collapse and recovery, we must decide which dimension we want to inhabit. Will we continue down the path of **environmental destruction**, or will we take the necessary steps to create a more **sustainable world**?

In the **Perceptual Multiverse**, the dimension we end up in is not predetermined—it is shaped by our actions, beliefs, and perceptions. If enough people commit to the reality of climate change and work toward solutions, we may be able to shift into the **dimension of recovery**. However, if we continue to ignore the warnings and prioritize short-term economic gain over long-term environmental health, we may find ourselves locked in the dimension of collapse.

The battle between these two dimensions—**collapse and recovery**—is one of the defining conflicts of our time. It is not just a battle over climate policy or environmental action—it is a battle over which reality we choose to inhabit. The future of the Earth, and the future of humanity, depends on which dimension we ultimately embrace.

Chapter 5: The Health Crisis Dimension

5.1 The Pandemic Divide: Health Science vs. Conspiracy

The Emergence of the Health Crisis Dimension

The **COVID-19 pandemic** of 2020 and beyond created one of the most intense flashpoints in modern history, revealing a stark divide between two major realities: those rooted in **health science** and those shaped by **conspiracy theories** and skepticism. The pandemic brought to light how deeply people's perceptions of **health, science**, and **authority** can differ, creating entirely separate **thought dimensions** within the broader society.

In the health science dimension, the COVID-19 pandemic is understood through the lens of **virology, epidemiology**, and public health policy. Scientists, healthcare professionals, and those who trust medical institutions accepted the overwhelming evidence that COVID-19 was a highly contagious virus, requiring global cooperation to prevent its spread. In this dimension, **mask mandates**, **social distancing**, and **vaccination** campaigns were seen as critical tools for protecting public health.

In contrast, a parallel dimension emerged where the pandemic was viewed with suspicion. In this dimension, COVID-19 was not seen as an existential threat, but rather as a **manipulated event**, a tool for **control**, or even a **hoax** designed to undermine freedoms. This reality was driven by **misinformation**, a deep-seated distrust of government and global institutions, and alternative theories about the virus's origin and purpose. Here, **conspiracy theories** flourished, framing the pandemic as part of a broader agenda.

These two dimensions were not merely different interpretations of the same events. They represented fundamentally different **realities** shaped by opposing beliefs about **science, authority,** and the nature of the pandemic itself. In this sense, the pandemic served as a flashpoint where **parallel realities** collided, making it clear that people were not simply disagreeing—they were inhabiting different dimensions of thought.

Health Science: Trust in Data and Expertise

In the dimension grounded in **health science**, the pandemic was a **biological crisis** that demanded an evidence-based response. From the earliest days of the outbreak, health experts turned to **data** and **models** to understand the virus's behavior and predict its spread. This dimension was built on trust in the **scientific method**—the idea that by collecting data, conducting experiments, and analyzing the results, humanity could find ways to control the pandemic and eventually defeat it.

For those living in this dimension, trust in institutions like the **World Health Organization (WHO)**, the **Centers for Disease Control and Prevention (CDC)**, and **national health agencies** was paramount. Public health guidelines, based on the best available evidence, were followed as part of a collective effort to protect society from the virus. This dimension emphasized the need for **global cooperation** and individual responsibility in the fight against COVID-19, with the understanding that the virus did not respect borders or political ideologies.

In this dimension, the focus was on **flattening the curve**—the idea that by slowing the spread of the virus, healthcare systems could avoid being overwhelmed, and fewer people would die. Measures such as **lockdowns**, **mask mandates**, and the eventual rollout of **vaccines** were seen as essential strategies for controlling the pandemic and returning to a semblance of normalcy. The overwhelming consensus within this dimension was that, while the measures were difficult and disruptive, they were necessary to save lives and prevent a more severe catastrophe.

Those who inhabited this dimension relied on the expertise of **virologists**, **epidemiologists**, and **healthcare workers** who had dedicated their careers to studying infectious diseases. These professionals became the guiding voices during the pandemic, and their recommendations were seen as the best way to navigate the crisis.

The Rise of the Conspiracy Dimension

Parallel to the health science dimension, another reality emerged where the pandemic was viewed through the lens of **conspiracy theories** and **skepticism**. In this dimension, the pandemic was not seen as a natural disaster or a public health emergency but as part of a broader **plot** to control or deceive the population. Those living in this dimension were often suspicious of government mandates, public health guidelines, and the **mainstream narrative** about the virus.

This dimension was shaped by **misinformation** spread through social media, alternative news sources, and online communities. **Conspiracy theories** about the virus's origin—whether it was a **bioweapon** engineered in a laboratory or a **hoax** designed to create fear—gained traction among individuals who were already predisposed to distrust authority. As the pandemic progressed, these theories evolved to include claims about **5G technology**, **Bill Gates**, and **population control**.

At the heart of the conspiracy dimension was a profound **mistrust** of the very institutions that were tasked with managing the pandemic. In this reality, the **WHO**, the **CDC**, and national governments were not seen as trustworthy sources of information, but as part of a **global elite** seeking to manipulate the population for political or financial gain. This dimension was fueled by a belief that the pandemic was being used as a pretext to erode personal freedoms, impose surveillance technologies, and restructure society.

For those inhabiting this dimension, measures like mask mandates and lockdowns were seen not as necessary public health interventions, but as **intrusions** on personal liberty. **Masks** were viewed as symbolic of government control, and resistance to wearing them became a form of protest against what was perceived as an authoritarian overreach. Similarly, **lockdowns** were framed as a way to cripple small businesses, destroy livelihoods, and increase dependence on the state.

This dimension was also characterized by a **rejection of expertise**. Health professionals, scientists, and even healthcare workers were viewed with suspicion. Instead of trusting data and evidence, individuals in this dimension often turned to **alternative sources of information**, such as YouTube channels, fringe websites, or figures who positioned themselves as outsiders challenging the "official narrative." These alternative sources often promoted **disinformation** that painted a starkly different picture of the pandemic than the one accepted by the health science dimension.

The Deepening Divide Between Health Science and Conspiracy

The clash between these two dimensions became one of the defining features of the pandemic, as each side viewed the other not just as misinformed, but as fundamentally **dangerous**. Those in the health science dimension saw conspiracy theories as a **threat to public safety**, arguing that misinformation about the virus was leading to unnecessary deaths and prolonging the pandemic. Public health officials warned that if people refused to follow guidelines—such as wearing masks, getting vaccinated, or adhering to social distancing—the virus would continue to spread, causing more harm.

On the other side, those in the conspiracy dimension viewed the health science community as complicit in a **global deception**. They believed that the measures being implemented were not about protecting public health, but about consolidating power and control. For them, the pandemic was an opportunity for governments and corporations to **reshape society** in ways that would benefit the elites while stripping away the freedoms of ordinary people.

This divide was exacerbated by **political polarization**, as the pandemic became entangled with broader cultural and ideological battles. In many countries, the response to COVID-19 became a **litmus test** for political identity, with positions on issues like masks, vaccines, and lockdowns aligning with broader political beliefs about government intervention, individual rights, and the role of science in society.

5.2 Vaccines, Misinformation, and Parallel Perceptions of Risk

Vaccines as a Flashpoint

One of the most contentious issues to arise during the pandemic was the development and distribution of **COVID-19 vaccines**. In the health science dimension, vaccines were seen as the key to ending the pandemic. The rapid development of vaccines, made possible by decades of prior research into mRNA technology and other vaccine platforms, was hailed as a scientific triumph. Vaccination campaigns were launched across the globe, with the hope that achieving **herd immunity** would prevent further outbreaks and allow societies to return to normal.

In this dimension, vaccines were viewed as a **public good**—a way for individuals to protect not only themselves but also their communities. Public health campaigns emphasized the importance of getting vaccinated to protect vulnerable populations, such as the elderly and those with underlying health conditions. For those in this dimension, the **risk-benefit analysis** was clear: the benefits of vaccination far outweighed any potential risks, and widespread vaccination was the most effective way to bring the pandemic to an end.

However, in the parallel dimension of **vaccine skepticism**, the rollout of COVID-19 vaccines was met with deep suspicion and resistance. In this dimension, vaccines were not seen as a solution to the pandemic, but as part of a broader **conspiracy** to manipulate or harm the population. The rapid development of the vaccines was viewed not as a scientific achievement, but as a **rushed process** that raised questions about safety and efficacy.

The Spread of Misinformation About Vaccines

Misinformation about vaccines played a central role in shaping this skeptical dimension. On social media platforms, alternative news sites, and fringe communities, a wide range of **false claims** about the COVID-19 vaccines proliferated. Some of these claims were rooted in existing vaccine skepticism, such as the belief that vaccines cause **autism** or other health problems. Others were specific to the COVID-19 vaccines, such as the false claim that they contained **microchips** or were designed to alter human DNA.

These conspiracy theories were often accompanied by **anecdotal evidence**—videos or stories from individuals who claimed to have suffered severe side effects after receiving the vaccine. While these stories were often unverified or taken out of context, they contributed to a growing sense of fear and mistrust among those who were already skeptical of vaccines. In this dimension, the idea of being **forced to take a vaccine** was seen as a violation of personal autonomy and bodily integrity.

The **misinformation ecosystem** around vaccines was vast and complex, with content being shared and reshared across platforms, creating an echo chamber where false claims were constantly reinforced. For those in the skeptical dimension, these alternative sources of information were more trustworthy than official channels. As a result, efforts by governments and public health organizations to promote vaccination were often met with resistance, with many viewing such campaigns as attempts to **coerce** or **deceive** the public.

Parallel Perceptions of Risk

One of the key differences between the health science dimension and the skeptical dimension was how each group perceived the **risks** associated with COVID-19 and the vaccines. In the health science dimension, the primary risk was the virus itself. COVID-19 was seen as a serious threat, particularly to older adults and individuals with pre-existing conditions. The risk of severe illness, hospitalization, or death from COVID-19 was considered far greater than the potential risks associated with the vaccines.

In the skeptical dimension, however, the **perceived risks** were inverted. For many in this dimension, the risks associated with the virus were downplayed, while the risks associated with the vaccine were magnified. This perception was often fueled by a belief that the virus was either less severe than reported or that the numbers of cases and deaths were being exaggerated. At the same time, the potential side effects of the vaccine were seen as far more dangerous than they actually were, with many individuals convinced that the vaccine could cause serious harm or even death.

These **parallel perceptions of risk** created a deep divide in how people responded to the pandemic. In the health science dimension, the emphasis was on **collective action**—getting vaccinated to protect oneself and others. In the skeptical dimension, the focus was on **individual autonomy**—rejecting vaccination in favor of personal freedom and a distrust of government mandates.

This divide was further complicated by the way **misinformation** shaped perceptions of the **vaccine's efficacy**. For example, when breakthrough cases of COVID-19 occurred in vaccinated individuals, conspiracy theorists seized on these cases as "proof" that the vaccine did not work, even though health experts explained that no vaccine is 100% effective and that the vaccines significantly reduced the severity of illness. This selective interpretation of data reinforced the parallel reality in which vaccines were seen as ineffective or even harmful.

5.3 How Flashpoints Shape the Health Crisis Reality

Misinformation and the Creation of Parallel Realities

One of the most important dynamics in the health crisis dimension is the role of **misinformation** in shaping parallel realities. During the COVID-19 pandemic, misinformation spread rapidly, fueled by social media algorithms that promoted **controversial** and **sensational** content. In many ways, the pandemic created a perfect storm for the proliferation of **conspiracy theories** and **misleading narratives**, as fear, uncertainty, and political polarization combined to create an environment where misinformation could thrive.

For those who inhabited the skeptical dimension, misinformation provided a coherent narrative that explained their fears and suspicions about the pandemic. The conspiracy theories that flourished in this dimension offered a sense of **certainty** and **control** in a time of chaos, providing a clear explanation for what was happening and who was to blame. In this way, misinformation acted as a kind of **dimensional anchor**, keeping individuals firmly rooted in a reality where the pandemic was not a health crisis, but a political and cultural battle for control.

At the same time, those in the health science dimension struggled to counter the spread of misinformation. Public health officials, scientists, and fact-checkers worked tirelessly to correct false claims and provide accurate information, but they were often hampered by the sheer volume of **disinformation** being spread online. As a result, the two dimensions became increasingly entrenched, with each side viewing the other as misguided or dangerous.

The Role of Flashpoints in Shaping Reality

Flashpoint issues like **mask mandates, vaccination requirements**, and **lockdowns** became central battlegrounds in the health crisis dimension, shaping the way people experienced the pandemic. Each flashpoint represented a moment where the two dimensions collided, with each side interpreting the same events in radically different ways.

For example, mask mandates were seen as a necessary public health measure in the health science dimension, but in the skeptical dimension, they were viewed as a symbol of government overreach and control. Similarly, vaccination campaigns were hailed as a life-saving intervention by those in the health science dimension, while those in the skeptical dimension saw them as part of a broader agenda to manipulate or harm the population.

These flashpoints did more than just highlight the divide between the two dimensions—they **shaped reality** for those who experienced them. In the health science dimension, each new mandate, policy, or scientific breakthrough was seen as part of the ongoing effort to control the pandemic and protect lives. In the skeptical dimension, each new development was interpreted as further evidence of the conspiracy at play, deepening the sense of mistrust and resistance.

Navigating the Health Crisis Dimension

Understanding the health crisis dimension through the lens of the **Perceptual Multiverse Theory** allows us to see how different realities can coexist within the same physical space, yet be completely separate in terms of perception and experience. The pandemic revealed just how difficult it can be to bridge the gap between these dimensions, as each side was deeply entrenched in its own reality, shaped by its own set of beliefs, values, and information sources.

For those in the health science dimension, the challenge was how to **communicate** with individuals in the skeptical dimension without alienating or antagonizing them. Efforts to promote public health measures were often met with resistance, not because the science was wrong, but because the individuals being targeted were inhabiting a different **dimensional framework**. In this sense, the pandemic highlighted the limitations of traditional methods of persuasion and the need for new approaches that take into account the existence of parallel realities.

At the same time, those in the skeptical dimension were often resistant to engaging with information that challenged their beliefs. The more entrenched they became in their reality, the harder it was to accept data or arguments from the health science dimension, leading to a **feedback loop** where each side became more convinced of its own truth.

Conclusion: The Legacy of the Health Crisis Dimension

The **COVID-19 pandemic** was not just a health crisis—it was a **dimensional crisis**, revealing the deep divisions in how people perceive and experience reality. The clash between the health science dimension and the skeptical dimension created intense flashpoints that shaped the way individuals and societies responded to the pandemic. These flashpoints were not just disagreements over policy or science—they were **manifestations** of parallel realities that existed within the same physical space.

As we move forward, understanding the dynamics of the health crisis dimension can help us navigate future challenges, whether they are related to public health, politics, or any other flashpoint issue. The **Perceptual Multiverse** reminds us that people can live in vastly different realities, shaped by their beliefs, experiences, and the information they consume. Recognizing this can help us find ways to **communicate** across these divides and work toward a more **unified** understanding of the world.

Chapter 6: Gender and Identity in the Multiverse

6.1 Fluid Dimensions: Gender as a Social Construct

The Evolution of Gender as a Fluid Concept

In recent decades, the conversation surrounding **gender** has expanded beyond the traditional binary framework of male and female. Within certain dimensions of thought and social understanding, gender is seen not as a fixed biological reality but as a **fluid, socially constructed identity**. This understanding of gender is rooted in the idea that the categories of "man" and "woman" are not determined by biology alone but are shaped by **cultural, historical, and social forces**.

In these **fluid dimensions**, gender is understood as something that exists along a spectrum, where individuals can identify anywhere between or beyond the traditional categories of male and female. This concept challenges the idea that gender is inherently tied to biological sex and emphasizes the **subjective experience** of gender identity. Those who exist within this dimension believe that gender is not something fixed at birth but is instead a personal and evolving expression of self.

This fluid understanding of gender has gained widespread recognition, particularly in **queer theory** and **feminist thought**. Scholars in these fields argue that rigid definitions of gender are socially constructed tools of control, used historically to maintain **patriarchal** systems and limit personal freedom. In this dimension, the idea of **non-binary** identities, **gender fluidity**, and **gender expression** as separate from biological sex are central to the conversation about how individuals define and experience their gender.

In the fluid dimension, **self-identification** is seen as a powerful and legitimate means of defining one's gender. Terms like **genderqueer**, **genderfluid**, **agender**, and **non-binary** have emerged to describe identities that exist outside the male-female binary. The belief here is that individuals have the right to define their own gender identity based on their internal experience, regardless of their physical or biological characteristics. For many in this dimension, gender is a deeply **personal** and **subjective experience**, shaped by one's own understanding of self rather than external societal expectations.

Historical and Cultural Contexts of Gender Fluidity

While the fluid dimension might seem like a modern development, the idea that gender is not fixed has deep historical and cultural roots. Many **Indigenous cultures** across the world have long recognized more than two genders. For example, in some Native American cultures, individuals who embody both masculine and feminine traits are known as **Two-Spirit people**, and they are often revered for their unique perspective and roles within the community. Similarly, in **South Asian cultures**, the **hijra** community has existed for centuries, embodying a gender identity that is neither strictly male nor female.

These examples highlight the fact that the **fluidity of gender** is not a new concept—it is simply one that has been marginalized or ignored in certain cultures, particularly in the **Western world**, where rigid binary definitions of gender have historically dominated. The resurgence of gender fluidity in contemporary discourse can be seen as a return to a broader, more inclusive understanding of gender that acknowledges the diversity of human experience.

In the **Perceptual Multiverse**, this fluid dimension of gender allows for a broad range of identities to coexist and be recognized. Here, gender is not seen as something one is born with, but as something one creates and navigates throughout their life. The boundaries of gender are flexible, and individuals can move between identities as they explore and discover their true selves.

The Role of Language in the Fluid Gender Dimension

One of the most important tools in the fluid dimension is **language**. The words we use to describe gender are powerful, and they shape how we understand and experience it. In the fluid dimension, there is a conscious effort to create and adopt language that reflects the diversity of gender identities. Terms like **they/them** pronouns, **ze/hir**, and **Mx.** have been introduced as alternatives to the traditional gendered language of **he/him** and **she/her**.

This emphasis on inclusive language serves two important functions. First, it allows individuals to express their identity in a way that feels authentic to them. For many people, using gender-neutral or non-binary pronouns is a way of asserting their place within the fluid dimension and rejecting the constraints of the binary system. Second, it challenges the broader society to rethink its assumptions about gender. By normalizing non-binary and gender-neutral language, the fluid dimension pushes for a more inclusive world where everyone's identity is recognized and respected.

However, the introduction of new language around gender is also a source of tension between the fluid dimension and the more **fixed dimensions** of gender, as discussed later in this chapter. Those who exist in fixed dimensions may feel uncomfortable or confused by the changing language, seeing it as unnecessary or even threatening to their understanding of the world.

The Political and Social Impact of Gender Fluidity

The recognition of gender as a social construct has had a profound impact on both **policy** and **social movements**. The rise of transgender and non-binary activism has pushed for legal changes to ensure that individuals can have their gender identities recognized on official documents, such as **passports**, **driver's licenses**, and **birth certificates**. Laws that protect against **discrimination based on gender identity** in the workplace, housing, and healthcare are key victories for those who inhabit the fluid dimension.

This movement toward greater acceptance of gender fluidity is also reflected in pop culture and media. Television shows, movies, and music increasingly feature **non-binary characters** and transgender individuals, helping to normalize gender diversity in the eyes of the broader public. For those in the fluid dimension, this visibility is crucial in the fight for recognition and equality.

However, the fluid dimension is not without its challenges. The rapid pace of change in how gender is understood and expressed has led to backlash from those who feel that the traditional gender system is under attack. The fluid dimension is often met with resistance from more **conservative or traditional dimensions** of gender, leading to flashpoints over issues like **bathroom access**, **transgender participation in sports**, and the role of gender in education.

6.2 Fixed Dimensions: Binary Gender and Biological Determinism

The Biological View of Gender as Fixed

In contrast to the fluid dimension, there are other dimensions where gender is seen as **binary** and fundamentally tied to biological sex. This dimension is rooted in the belief that gender is determined by **biological characteristics**, such as chromosomes, reproductive organs, and secondary sexual traits. In this dimension, individuals are classified as either male or female based on these physical characteristics, and this classification is considered fixed and immutable.

For those who inhabit this **fixed dimension**, gender is seen as something that is **inherited** and not subject to personal choice. This dimension views gender as a natural and essential part of human biology, with each individual being born either male or female. From this perspective, gender roles and expectations are closely tied to biological sex, and there is little room for fluidity or ambiguity.

The idea of **gender as a binary** is deeply ingrained in many cultural, religious, and social traditions. In many societies, being male or female is not only a biological reality but also comes with specific **social roles** and **expectations**. Men are often expected to embody traits such as strength, leadership, and rationality, while women are expected to be nurturing, empathetic, and focused on caregiving. These roles are seen as natural extensions of biological differences and are reinforced through cultural norms and institutions.

The Role of Biological Determinism in Gender

At the heart of the fixed dimension is the concept of **biological determinism**—the idea that human behavior, traits, and roles are determined by biology. This view holds that men and women are inherently different due to their biological makeup, and these differences dictate how they should behave and what roles they should occupy in society. In this dimension, gender is seen as a direct reflection of sex, with biological differences driving the distinctions between masculinity and femininity.

For example, proponents of biological determinism might argue that men are naturally more aggressive and competitive because of **higher levels of testosterone**, while women are more nurturing due to their role in childbearing and breastfeeding. In this dimension, these biological differences are seen as fundamental and unchangeable, and they justify the existence of distinct gender roles in society.

In this reality, gender roles are often seen as necessary for maintaining **social order**. Traditional families, with clearly defined roles for men and women, are viewed as the foundation of society, providing stability and continuity. Those who exist in this dimension may view challenges to the binary gender system—such as the rise of non-binary identities or transgender rights—as a threat to the natural order.

Religious and Cultural Roots of Binary Gender

The fixed dimension of binary gender is often closely tied to **religious and cultural beliefs**. In many religious traditions, gender is seen as divinely ordained. For example, in the **Abrahamic religions** (Judaism, Christianity, and Islam), the belief that God created humans as male and female is a central tenet. In these faiths, gender roles are often prescribed by religious texts, with men and women assigned specific responsibilities within the family and community.

In this dimension, gender is not just a biological reality—it is also a moral one. Gender roles are seen as part of a **divine plan**, and deviations from these roles are often viewed as unnatural or sinful. This can lead to significant resistance to the idea of gender fluidity or the recognition of transgender identities, as these concepts challenge deeply held religious beliefs about the nature of gender and human identity.

Culturally, many societies have long upheld the binary gender system as a way of organizing social life. The division of labor, the structure of families, and the distribution of power have all been shaped by the belief that men and women are fundamentally different. In these cultures, gender roles are passed down from generation to generation, and they are seen as essential to the preservation of **tradition** and **cultural continuity**.

The Resistance to Gender Fluidity

For those who inhabit the fixed dimension, the rise of gender fluidity is often seen as a **direct challenge** to the natural and moral order. The idea that gender can be self-identified or that people can exist outside the male-female binary is often met with confusion, discomfort, or outright rejection. This resistance is not just about a disagreement over gender identity—it is about a fundamental clash of dimensions, where the very nature of gender is understood in radically different ways.

In this dimension, there is a strong emphasis on **biological facts** as the basis for understanding gender. Individuals who advocate for the fixed dimension often argue that while people may feel differently about their gender identity, these feelings do not change the biological reality of their sex. They may view efforts to recognize non-binary or transgender identities as a form of **social engineering** that undermines the importance of biological distinctions.

Moreover, the fixed dimension often views the changes occurring in society—such as the increasing recognition of non-binary identities, the push for transgender rights, and the growing acceptance of diverse gender expressions—as part of a broader **cultural shift** that threatens traditional values. For many, these changes are seen as indicative of a society that is losing touch with its roots, abandoning the clear distinctions between men and women that have long been considered essential to social order.

6.3 The Intersection of Dimensions: Where Identity Becomes Reality

Navigating the Tension Between Fluid and Fixed Dimensions

The clash between the fluid and fixed dimensions of gender is one of the most intense **flashpoints** in contemporary society. Each dimension operates with its own set of assumptions, values, and understandings about the nature of gender, making it difficult for people from these dimensions to find common ground. However, in the **Perceptual Multiverse**, these dimensions often intersect, leading to moments where individuals are forced to confront the existence of multiple, conflicting realities.

For example, the debate over **transgender rights** often serves as a flashpoint where the fluid and fixed dimensions collide. In the fluid dimension, the recognition of transgender identities is seen as a matter of human rights and personal autonomy. People should be free to define their own gender and have that identity respected by society, regardless of their biological sex. This dimension emphasizes the importance of **affirming gender identities** and providing legal protections for those who do not fit within the traditional binary system.

In contrast, those in the fixed dimension may see the recognition of transgender identities as a threat to the natural order and the integrity of biological distinctions. For them, the idea that someone can change their gender or exist outside the binary is fundamentally at odds with the reality of biological sex. This dimension often frames the debate over transgender rights as a conflict between **truth** and **ideology**, with biological facts being pitted against what they see as a socially constructed and politically motivated narrative.

Identity as a Construct and Reality as a Choice

One of the central questions raised by the intersection of these dimensions is whether **identity is a construct** or a **biological reality**. In the fluid dimension, identity is seen as something that is constructed through **personal experience**, **self-discovery**, and **social interaction**. People are not born with a fixed identity—they create it over time, based on their internal sense of self and their interactions with the world around them. This dimension places a strong emphasis on the **subjectivity of identity**—the idea that only the individual can truly know who they are.

In the fixed dimension, identity is seen as something that is **given at birth** and determined by biological characteristics. From this perspective, identity is not something that can be changed or constructed—it is something that is inherent and objective. The belief here is that people are born with a specific identity (male or female), and this identity is fixed by nature.

At the intersection of these dimensions, the question of whether identity is **fluid** or **fixed** becomes a matter of reality itself. For those in the fluid dimension, the ability to self-identify and express one's gender in diverse ways is a fundamental aspect of human freedom. For those in the fixed dimension, the insistence on fluidity represents a rejection of biological truths and a departure from the natural order.

The Role of Legal and Institutional Recognition in Shaping Reality

One of the key battlegrounds at the intersection of these dimensions is the question of **legal and institutional recognition**. As more people identify as non-binary or transgender, there has been a push for **legal recognition** of gender identities beyond the male-female binary. In many countries, individuals can now change the gender marker on their official documents, and some legal systems recognize non-binary as a legitimate gender category.

For those in the fluid dimension, these changes represent a significant step forward in the recognition of **gender diversity** and the validation of individual identities. However, for those in the fixed dimension, these legal changes are often seen as a dangerous departure from the truth of biological sex. This dimension may view the recognition of non-binary or transgender identities as an erosion of traditional values and a threat to the integrity of legal and social systems.

The question of **bathroom access**, for example, has become a flashpoint in many countries. In the fluid dimension, allowing people to use the bathroom that corresponds to their gender identity is seen as a matter of basic human dignity and respect. In the fixed dimension, this policy is often viewed as a violation of privacy and safety, particularly for women and girls. Each dimension interprets the issue through its own lens, making it difficult to find a solution that satisfies both perspectives.

Conclusion: The Future of Gender and Identity in the Multiverse

As society continues to grapple with the complex questions surrounding gender and identity, it is clear that the **Perceptual Multiverse** offers a useful framework for understanding the multiple dimensions at play. The clash between the fluid and fixed dimensions of gender is not just a disagreement over facts or policies—it is a fundamental clash of realities, where each dimension operates with its own set of beliefs, values, and truths.

In the coming years, the intersection of these dimensions will likely continue to be a source of tension and conflict, particularly as the push for greater recognition of **gender diversity** gains momentum. However, the Perceptual Multiverse also offers the possibility of finding ways to navigate these **competing dimensions** and create a more inclusive world where multiple realities can coexist.

Ultimately, the future of gender and identity in the multiverse will depend on our ability to recognize the legitimacy of each dimension while working to build bridges between them. Whether we are able to move toward a more fluid understanding of gender or whether the binary system remains dominant will shape not only individual experiences but also the broader social and legal frameworks that govern our lives.

Chapter 7: Free Speech vs. Hate Speech

7.1 Objective Reality: Free Speech as a Fundamental Right

The Foundations of Free Speech

In many democratic countries, **free speech** is enshrined as a fundamental right and cornerstone of democracy. In the United States, for instance, the **First Amendment** of the Constitution guarantees the right to free expression, including the freedom of speech, press, and assembly. Similarly, in other nations such as **Canada**, **Germany**, and the **United Kingdom**, free speech is protected under laws that ensure individuals can express their views without fear of government censorship or reprisal.

The rationale behind protecting free speech is rooted in the belief that open discourse is essential for the functioning of a healthy democracy. The **marketplace of ideas** theory suggests that when people are free to share their thoughts and opinions, the best ideas will naturally rise to the surface through debate and discussion. Moreover, free speech allows individuals to hold governments and institutions accountable, ensuring that power is not abused, and that citizens can challenge policies and norms.

This view of free speech as a **universal good** is often linked to the idea that expression, in all its forms, is a fundamental part of **human autonomy**. In this dimension, the freedom to speak one's mind is seen as an extension of the individual's right to self-determination and personal liberty. When governments or institutions attempt to limit speech, they are perceived as encroaching on the individual's most basic rights.

Legal Protections and Limitations

However, even in societies that prioritize free speech, there are often legal limitations. Speech that directly incites **violence**, poses a threat to **national security**, or spreads **defamation** is not protected under most legal systems. For example, the U.S. Supreme Court has established certain categories of speech—such as **true threats**, **obscenity**, and **fighting words**—that are not protected by the First Amendment. These exceptions are intended to prevent speech that could cause direct harm or violate the rights of others.

In other countries, the balance between **free speech** and **social responsibility** is more heavily weighted in favor of restricting certain kinds of harmful expression. For instance, many European countries have laws against **Holocaust denial** or the glorification of **Nazi symbols**, reflecting the belief that such speech poses a unique threat to public order and the dignity of those affected by historical atrocities. Similarly, **Germany's** Basic Law protects free speech, but Article 5 of the Basic Law includes limits, particularly regarding speech that incites hatred against people based on their race, religion, or ethnicity.

In this **objective reality**, the tension between protecting **free expression** and safeguarding public welfare is a recurring theme. The **legal frameworks** that regulate speech vary widely across the world, reflecting different cultural attitudes toward the balance between individual liberty and collective safety. But even in countries where free speech is a cherished right, there are ongoing debates about how far this protection should extend and where the line should be drawn.

7.2 Subjective Perception: Free Speech Without Limits vs. Restricting Harmful Speech

The "No Limits" Dimension: Absolute Freedom of Expression

In one dimension of the **free speech debate**, individuals believe that **free expression should have no limits**. This perspective is rooted in the idea that any restriction on speech—no matter how harmful it might seem—sets a dangerous precedent for **censorship** and undermines the principle of individual liberty.

For those who exist in this dimension, free speech is an absolute right, and any attempt to regulate or restrict it is viewed as an assault on **personal autonomy** and a step toward authoritarianism. From this point of view, even offensive, harmful, or misleading speech must be protected because the alternative—allowing governments or institutions to decide what speech is acceptable—could lead to the suppression of dissenting opinions.

This dimension of the free speech debate emphasizes the importance of **tolerating unpopular or controversial speech** in order to preserve a truly open society. As the philosopher **John Stuart Mill** argued in his essay *On Liberty*, even ideas that are considered false or harmful can contribute to the pursuit of truth, because they force society to **re-examine** its assumptions and beliefs. From this perspective, banning or restricting certain types of speech may prevent people from discovering new truths or challenging outdated or unjust norms.

The **"no limits"** dimension also highlights the **slippery slope** argument, which suggests that once a society begins to restrict certain kinds of speech, it becomes easier for those in power to broaden the scope of censorship. What begins as a well-intentioned effort to ban **hate speech**, for example, could eventually expand to include restrictions on **political dissent**, **satire**, or even **artistic expression**. To avoid this, those in this dimension argue that **all speech**, no matter how controversial or offensive, must be protected.

The Case for Restricting Harmful Speech

In contrast, many people believe that there are certain kinds of speech that should be restricted because of the harm they can cause. For those who inhabit this dimension, free speech is not an absolute right but one that must be balanced against the need to protect vulnerable individuals and groups from **hate speech**, **misinformation**, or **incitement to violence**.

This dimension is grounded in the belief that speech is not merely an act of expression but also a **powerful tool** that can shape reality, influence behavior, and cause harm. Hate speech, for example, is seen not just as offensive language but as a form of **violence** that can lead to real-world consequences for the targeted groups. **Racist, sexist, homophobic**, or **transphobic speech** can contribute to a climate of **fear, discrimination,** and **exclusion**, making it harder for marginalized communities to participate fully in society.

In this dimension, the goal is to create a **balance** between protecting the right to free speech and protecting individuals from **harmful rhetoric**. Many advocates for this perspective argue that **hate speech laws** are necessary to prevent the escalation of violence and discrimination. For example, **hate speech** against minority groups can lead to **social exclusion**, economic marginalization, and even **hate crimes**. By restricting this kind of speech, societies can help create a safer, more inclusive environment for everyone.

This dimension also argues that **misinformation**—especially when it comes to issues like **public health** or **elections**—can have devastating consequences. During the COVID-19 pandemic, for example, the spread of **misinformation** about vaccines and public health measures led to **vaccine hesitancy**, which in turn prolonged the pandemic and resulted in avoidable deaths. In this reality, restricting the spread of false or misleading information is seen as essential to protecting public health and democracy.

The Role of Power Dynamics in the Free Speech Debate

An important element of the **free speech vs. hate speech** debate is the role of **power dynamics**. Those in favor of restricting harmful speech often argue that **free speech protections** disproportionately benefit those who are already in positions of power. For example, **hate speech** laws are designed to protect **vulnerable communities**—such as racial minorities, LGBTQ+ individuals, and religious groups—who may be disproportionately affected by hate speech, while those in positions of power (such as politicians, media figures, or corporate leaders) are less likely to face the same kinds of consequences for their speech.

This dimension also emphasizes the idea that **speech is not neutral**—it reflects and reinforces existing power structures. Hate speech can perpetuate **structural inequality**, just as

misinformation can undermine public trust in institutions or erode the foundations of democracy. From this perspective, allowing harmful speech to go unchecked can contribute to the marginalization of vulnerable groups and exacerbate social divides.

7.3 How Flashpoints Shape the Free Speech Reality

Flashpoint Issues: The Intersection of Free Speech and Hate Speech

The tension between **free speech** and **hate speech** has become one of the defining flashpoints in contemporary society. Flashpoint issues—such as the use of **racial slurs**, **offensive symbols**, or **misinformation**—often reveal the deep divide between those who view free speech as an absolute right and those who believe it must be restricted to protect vulnerable groups.

One example of this flashpoint is the debate over **online speech**. Social media platforms have become a battleground for free speech, with companies like **Facebook**, **Twitter**, and **YouTube** facing criticism for either allowing too much harmful content or for restricting speech in ways that are perceived as biased or unfair. In the wake of controversies such as the **Capitol riots** in the United States or **hate speech** against ethnic minorities in Myanmar, social media companies have come under pressure to remove harmful content and prevent their platforms from being used to incite violence. At the same time, critics argue that these companies have too much power to decide what is and isn't acceptable speech, raising concerns about **corporate censorship**.

Another flashpoint issue is the debate over **free speech on college campuses**. Many universities have become arenas for the clash between **free expression** and the desire to create safe and inclusive environments for students. Some argue that **trigger warnings** and **safe spaces** are necessary to protect students from harmful or traumatic speech, while others contend that these measures are stifling free debate and intellectual diversity.

In both cases, the tension between free speech and hate speech reflects a broader **cultural divide** over the meaning and limits of free expression. On one side are those who believe that protecting **free speech**—even when it is offensive or controversial—is essential to the functioning of a free society. On the other side are those who argue that certain types of speech are so harmful that they must be restricted to ensure the safety and dignity of all individuals.

The Role of Social Media and the Internet in Amplifying Flashpoints

Social media platforms play a crucial role in amplifying the **free speech vs. hate speech** debate, creating new flashpoints and intensifying the divide between different dimensions of reality. Platforms like **Facebook**, **Twitter**, and **Reddit** allow for the rapid dissemination of both **speech** and **misinformation**, often blurring the line between personal expression and public harm.

One of the defining features of social media is that it provides a platform for individuals to speak to large audiences, often without the **editorial oversight** that traditional media outlets provide. This has led to an explosion of **unregulated speech**, where **misinformation, hate speech**, and **conspiracy theories** can spread unchecked. In the health dimension, for example, **COVID-**

19 **misinformation** spread rapidly online, leading to vaccine hesitancy and public confusion about the severity of the virus.

At the same time, social media platforms have become targets of **public scrutiny** for their role in regulating speech. Platforms have adopted various approaches to **content moderation**, including banning certain users, removing harmful content, and labeling misleading posts. However, these measures have sparked a fierce debate over whether social media companies should have the power to regulate speech, with critics arguing that this constitutes **corporate censorship** and threatens the open exchange of ideas.

The role of social media in shaping the **free speech vs. hate speech** debate is further complicated by the **global nature** of these platforms. What is considered hate speech in one country might be protected speech in another, leading to inconsistencies in how content is moderated across different cultural and legal contexts. This creates a challenging dynamic, where individuals from different dimensions of reality come into contact, often leading to heated flashpoints over issues of free expression and social responsibility.

7.4 Navigating the Tension Between Free Speech and Hate Speech

The Importance of Context in Free Speech Debates

One of the key challenges in navigating the tension between free speech and hate speech is understanding the **context** in which speech occurs. Speech that may seem harmless or benign in one context can take on a much more harmful dimension when it is directed at a marginalized or vulnerable group. For example, a racial slur used in a private conversation may be seen as offensive but relatively inconsequential, while the same slur used in a **public rally** or on social media can contribute to a climate of fear and discrimination.

In this sense, **context matters** when evaluating whether certain types of speech should be restricted. Laws that prohibit hate speech, for example, often take into account the **historical and social context** of the speech in question. In countries like **Germany** and **France**, Holocaust denial is not just considered offensive—it is seen as a form of hate speech that threatens the **collective memory** of historical atrocities and the dignity of the victims. Similarly, laws against incitement to violence take into account the **potential impact** of speech on public order and safety.

At the same time, those who advocate for **unrestricted free speech** argue that context should not determine whether speech is protected. From this perspective, individuals should be free to express their views—no matter how controversial or offensive—without fear of legal consequences or social reprisal. In this view, restricting speech based on context can lead to a **slippery slope** where more and more speech is censored in the name of protecting vulnerable groups.

Finding a Balance: Protecting Free Expression and Vulnerable Groups

Finding a balance between protecting **free expression** and protecting vulnerable groups from harm is one of the most challenging aspects of the free speech vs. hate speech debate. On one hand, free speech is essential for ensuring that all voices can be heard and that ideas can be debated in the open. On the other hand, certain kinds of speech can cause real harm to individuals and communities, contributing to a climate of fear, discrimination, and exclusion.

Many legal systems attempt to strike a **balance** between these competing interests by focusing on the **intent** and **impact** of speech. For example, speech that is intended to incite violence or promote hatred against a particular group is often considered beyond the bounds of free expression, as it poses a direct threat to public safety and the well-being of vulnerable communities. In contrast, speech that is simply **offensive** or **controversial**—but does not incite violence or cause direct harm—may be protected under free speech laws, even if it is unpopular.

Ultimately, the balance between free speech and hate speech depends on how a society prioritizes the values of **individual freedom** and **social responsibility**. In some societies, free speech is considered the highest value, and restrictions on speech are seen as a threat to **democratic discourse**. In others, the protection of vulnerable groups from harmful speech is considered equally important, and hate speech laws are viewed as necessary to promote **social harmony** and protect human dignity.

Conclusion: The Future of Free Speech in the Multiverse

The debate over **free speech vs. hate speech** is not just about legal or philosophical principles—it is about the **tension between competing realities**. In one dimension, free speech is seen as an absolute right, essential for the functioning of a free society. In another, the protection of vulnerable groups from harmful speech is prioritized, and free speech is seen as a right that comes with **responsibilities**.

As we navigate the future of **free speech** in the **Perceptual Multiverse**, we must recognize that these dimensions will continue to intersect and clash. The challenge is to find ways to protect the **fundamental right to free expression** while also ensuring that speech is not used as a weapon to **harm, divide, or oppress**.

The future of the free speech vs. hate speech debate will likely be shaped by the **evolution of technology**, the **changing nature of public discourse**, and the ongoing tension between **individual liberty** and **social responsibility**. As we move forward, we must grapple with the question of how to create a society where all voices can be heard, but where no one is silenced or marginalized by harmful speech.

Chapter 8: The Sanctity of Life Dimension

8.1 Abortion and the Dividing Line of Reality

The Clash of Realities: Two Irreconcilable Dimensions

Abortion has long been one of the most contentious and polarizing issues in societies across the world. This divide is not simply a matter of differing opinions on a single issue; it represents a **clash of two fundamentally different realities**—what can be understood as separate **dimensions of belief**. On one side of this divide is a reality where the **right to bodily autonomy** is seen as paramount, and access to abortion is regarded as an essential part of a person's freedom to make decisions about their own body. In this dimension, the focus is on **individual choice** and **reproductive rights**.

On the other side is a dimension where the **sanctity of life**—specifically the life of the unborn—is seen as the highest moral principle. Here, abortion is viewed as the taking of a human life, and the fetus is considered a person with the same moral rights as any other human being. In this dimension, the primary focus is on **protecting the innocent** and ensuring the right to life is extended to the unborn.

The **dividing line** between these two realities is stark, and because the moral, ethical, and emotional stakes are so high, bridging the gap between these dimensions can seem impossible. Each side perceives the other as being fundamentally wrong, not only about abortion but about the underlying principles that guide their beliefs.

Historical Context: How Abortion Became a Flashpoint

The **abortion debate** is deeply rooted in **historical, cultural, and religious contexts**. Historically, many societies had strict prohibitions against abortion, often grounded in religious doctrines that emphasized the sanctity of life from the moment of conception. In Christian theology, for instance, the belief that life begins at conception has played a major role in shaping anti-abortion views, particularly in **Catholicism** and **Evangelical Christianity**. In these traditions, abortion is seen as a grave moral offense—a violation of divine law.

In the mid-20th century, however, many countries began to re-evaluate their stance on abortion as part of broader movements toward **women's rights** and **gender equality**. In 1973, the landmark **Roe v. Wade** decision in the United States made abortion a constitutionally protected right, and other nations followed suit by liberalizing their abortion laws. In this new dimension, access to safe and legal abortion became seen as a vital part of **reproductive freedom** and a woman's ability to control her own destiny.

Yet, even as legal protections for abortion expanded, the opposition to abortion remained strong, particularly among religious groups. For those in the **sanctity of life dimension**, the legalization of abortion represented not a victory for freedom, but a profound moral crisis—a state-sanctioned violation of the most fundamental human right: the right to life.

The Abortion Debate as a Flashpoint Between Parallel Realities

The **abortion debate** is more than just a political or legal dispute—it is a **flashpoint** between parallel realities, where individuals from different dimensions perceive the world in entirely different ways. For those in the **bodily autonomy dimension**, abortion is not a moral issue in the same way it is for those in the **sanctity of life dimension**. Instead, it is viewed primarily through the lens of **individual freedom, personal health**, and **reproductive justice**.

In the bodily autonomy dimension, the ability to access abortion is seen as a crucial part of one's **right to control their own body**. Those who inhabit this dimension often focus on the implications of **forced pregnancy** and how denying abortion can have devastating consequences for a person's health, economic stability, and future opportunities. The moral calculus here is centered on the well-being and rights of the individual carrying the pregnancy, rather than on the fetus.

Conversely, those in the **sanctity of life dimension** frame the issue of abortion as one of **moral absolutes**. For them, life begins at conception, and the unborn child is viewed as a person with inherent moral value. Abortion, therefore, is seen as the unjust taking of a human life, akin to **murder**. In this dimension, the rights of the fetus are paramount, and any discussion of abortion must begin with the assumption that the fetus has the same moral status as a born human being.

The **flashpoint** nature of the abortion debate is most evident in the way each side reacts to attempts to legislate or regulate abortion. Proponents of abortion rights see efforts to restrict access to abortion as an assault on personal freedom, while anti-abortion advocates view the legalization of abortion as a gross violation of human rights. These opposing perspectives are not easily reconciled because they are grounded in different **moral dimensions**—different realities where the meaning of life, freedom, and rights diverge.

8.2 Bodily Autonomy vs. The Right to Life: Two Parallel Universes

The Bodily Autonomy Dimension: The Right to Choose

In the **bodily autonomy dimension**, the central principle is that every person has the right to make decisions about their own body. This dimension is rooted in the belief that individual freedom extends to the most personal aspects of one's life, including decisions about **reproduction**. The argument here is that forcing someone to carry an unwanted pregnancy to term violates their basic human rights by denying them control over their own body.

This dimension places a high value on **personal autonomy** and **self-determination**. It argues that decisions about whether to continue or terminate a pregnancy are deeply personal and should be made by the individual affected, not by the government or society at large. For those in this dimension, the question is not about when life begins, but about who has the right to make decisions about a pregnancy.

Proponents of bodily autonomy argue that access to **safe and legal abortion** is essential for ensuring gender equality. They contend that without the ability to control their reproductive lives, individuals—particularly women—are at a disadvantage in terms of education, employment, and economic independence. The decision to have an abortion is framed as a **healthcare issue**, where the focus is on the individual's well-being and future opportunities.

In this dimension, the fetus is often seen as part of the pregnant person's body until viability, and thus, the rights and interests of the pregnant individual take precedence. This does not mean that those in this dimension do not value life, but rather that they see the right to bodily autonomy as being of greater moral importance in the early stages of pregnancy. The **right to choose** is seen as fundamental to living a life of dignity and freedom.

The Right to Life Dimension: Protecting the Unborn

In stark contrast, the **right to life dimension** centers on the belief that **life begins at conception**, and that from the moment of conception, the fetus is a human being with its own moral and legal rights. Those who inhabit this dimension believe that the right to life is the most fundamental of all rights, and that society has a moral obligation to protect the unborn.

For those in this dimension, the issue of abortion is framed not in terms of personal choice but in terms of **moral responsibility**. They argue that society must prioritize the protection of the most vulnerable members, including the unborn, who cannot speak for themselves. In this dimension, the moral status of the fetus is non-negotiable—it is considered a human being with the same intrinsic value as any other person.

The **pro-life movement**, which is rooted in this dimension, advocates for policies that **restrict or ban abortion**, often with the goal of overturning legal precedents like **Roe v. Wade** in the United States. For pro-life advocates, the legalization of abortion is seen as a grave injustice that must be corrected. In their view, allowing abortion to continue unchecked is equivalent to **sanctioning murder** at the societal level.

The right to life dimension is often informed by **religious beliefs** about the sanctity of life. Many individuals in this dimension are guided by the conviction that human life is sacred because it is created by God, and therefore, no one has the right to end it. This religious dimension adds a profound moral weight to the debate, as those who oppose abortion see it not only as a legal issue but as a **spiritual** and **moral** battle.

The Incommensurability of the Two Dimensions

One of the greatest challenges in the abortion debate is the **incommensurability** of the two dimensions—bodily autonomy and the right to life. These dimensions operate on **different moral frameworks** that do not easily overlap, making it difficult for individuals from either side to engage in meaningful dialogue.

In the bodily autonomy dimension, the **rights of the pregnant individual** are the central concern, and the fetus is not viewed as having the same moral status until a certain point in the pregnancy, such as **viability**. In this reality, discussions about the morality of abortion focus on the rights of the individual to make choices about their own life, without interference from the state or society.

In the right to life dimension, however, the **fetus** is viewed as a person from the moment of conception, and therefore has rights that must be protected. In this dimension, the focus is on the **moral duty** to defend the lives of the unborn, even if that means placing restrictions on the rights of the pregnant individual. These two perspectives are fundamentally at odds because they begin from different assumptions about the nature of life, personhood, and rights.

8.3 The Moral Battle Between Dimensions of Belief

Moral Absolutism vs. Moral Relativism

At the heart of the abortion debate is a broader conflict between two competing moral frameworks: **moral absolutism** and **moral relativism**. These frameworks shape how individuals in each dimension view the issue of abortion and inform their broader beliefs about morality, ethics, and the role of society.

In the **right to life dimension**, the moral framework is often one of **absolutism**. Here, the belief is that certain moral truths are universal and unchanging—namely, that life is sacred and must be protected at all costs. This dimension is often informed by **religious principles** or deeply held philosophical convictions about the inherent value of human life. Those who hold this view believe that abortion is morally wrong in all or most cases because it violates the absolute principle of the **sanctity of life**.

In contrast, the **bodily autonomy dimension** often operates on a framework of **moral relativism**, where the context of each situation matters, and ethical decisions must take into account individual circumstances. In this view, morality is not always black and white, and decisions about complex issues like abortion must consider factors such as **personal freedom**, **health risks**, and **social inequalities**. For those in this dimension, what is right or wrong depends on the situation, and the individual's autonomy plays a central role in determining the moral outcome.

This **clash of moral frameworks** contributes to the deep divisions between the two dimensions, as each side believes that the other is fundamentally mistaken about the nature of morality. Those who subscribe to moral absolutism often see relativism as dangerous, arguing that it leads to moral decay and the erosion of universal values. Conversely, those who adopt a relativist approach may view absolutism as rigid and unyielding, unable to accommodate the complexities of real-life situations.

Religious vs. Secular Dimensions

Another key factor in the abortion debate is the tension between **religious** and **secular** perspectives on morality and ethics. For many people in the right to life dimension, opposition to abortion is rooted in **religious beliefs** about the sanctity of life. In religious traditions such as **Christianity**, **Judaism**, and **Islam**, there is often a belief that human life is a gift from God and that taking a life—whether born or unborn—is morally wrong.

In these religious dimensions, the question of abortion is not just a legal or political issue—it is a **spiritual battle** between good and evil. Many religious individuals see their opposition to abortion as part of their duty to **uphold divine law** and protect the innocent. For them, legalizing abortion is a profound moral failure, and efforts to restrict or ban abortion are seen as a way of restoring **moral order** to society.

On the other hand, many individuals in the **bodily autonomy dimension** approach the issue from a **secular** perspective, where morality is not determined by religious doctrine but by considerations of **human rights**, **personal freedom**, and **ethical reasoning**. For those in this dimension, the decision to have an abortion is a personal one that should be based on the individual's circumstances and needs, without interference from religious or governmental authorities.

This tension between **religious** and **secular** dimensions is evident in the public discourse surrounding abortion. Religious groups often frame the debate in terms of moral absolutes and divine will, while secular advocates focus on issues of **gender equality**, **reproductive justice**, and the need for **separation of church and state**. These competing worldviews make it difficult to find common ground, as each side approaches the issue from fundamentally different assumptions about the role of religion in public life.

The Role of Empathy in the Abortion Debate

One of the most challenging aspects of the abortion debate is the lack of **empathy** between the two dimensions. Those who are firmly entrenched in either the bodily autonomy dimension or the right to life dimension often struggle to understand the motivations and emotions of those on the other side. This lack of empathy can lead to increased polarization and hostility, making it even more difficult to engage in meaningful dialogue.

For example, those in the right to life dimension may have difficulty empathizing with individuals who seek abortions, viewing them as morally irresponsible or selfish. They may not fully appreciate the emotional, physical, and economic burdens that can come with an unwanted pregnancy, particularly for those who lack access to healthcare, financial stability, or social support.

Conversely, those in the bodily autonomy dimension may struggle to empathize with individuals who oppose abortion on moral or religious grounds. They may see anti-abortion activists as trying to control women's bodies or impose their beliefs on others, without fully understanding the depth of conviction that drives the pro-life movement.

Building empathy between these dimensions is crucial for finding a way forward in the abortion debate. This requires recognizing that individuals on both sides of the issue are often motivated by deeply held beliefs and values, and that the **emotional and moral stakes** are high for everyone involved. Empathy can help bridge the divide by fostering a more nuanced and compassionate understanding of the complexities of abortion and the realities that shape people's lives.

Conclusion: The Ongoing Battle for the Sanctity of Life Dimension

The debate over abortion is not just a legal or political issue—it is a **clash of dimensions** that reflects deeper philosophical, moral, and cultural differences about the meaning of life, freedom, and human rights. In the **Perceptual Multiverse**, the bodily autonomy and right to life dimensions exist in parallel, each operating with its own set of moral principles and assumptions.

As societies continue to grapple with the question of abortion, the tension between these dimensions is likely to persist. Efforts to find common ground will require a willingness to engage with the underlying beliefs that shape each dimension and to recognize the legitimate concerns and values on both sides. The **future of the abortion debate** will depend on whether these dimensions can coexist in a way that respects both individual autonomy and the sanctity of life.

Ultimately, the abortion debate is a reflection of the broader struggle to navigate **competing realities** in a world where people hold profoundly different views about what it means to be human. By understanding the dimensions of belief that shape this debate, we can begin to find ways to bridge the divide and create a more compassionate and just society for all.

Chapter 9: Economic Inequality as Dimensional Conflict

9.1 Redistribution vs. Free Market: Two Economic Realities

The Free Market Dimension: The Promise of Capitalism

In the **free market dimension**, capitalism is seen as the most efficient and natural economic system, one that fosters innovation, rewards hard work, and creates opportunities for individuals to rise based on their merit. This dimension is grounded in the belief that the **free market**—when left to operate without heavy government interference—naturally leads to the greatest prosperity for the largest number of people.

In this dimension, economic success is viewed as a reflection of individual effort, talent, and perseverance. Those who achieve wealth are seen as deserving it, having worked hard to earn their success. **Competition** is considered a driving force behind **innovation** and **efficiency**, as it pushes individuals and companies to improve their products, services, and business practices. The **invisible hand** of the market is believed to guide resources to their most productive uses, creating wealth and improving the standard of living for society as a whole.

Advocates of the free market dimension believe that **minimal government regulation** is essential to maintaining the freedom and dynamism of the market. Too much government intervention—whether through taxes, regulations, or social welfare programs—distorts the natural workings of the market and creates **inefficiencies** that stifle growth. In this reality, the best way to combat poverty and inequality is not through redistribution but by promoting **economic growth** and ensuring that individuals have the freedom to pursue their own interests.

This dimension often emphasizes the importance of **personal responsibility** and **individualism**. People are seen as largely responsible for their own economic outcomes, and success is viewed as attainable for anyone willing to work hard and take risks. From this perspective, wealth inequality is not necessarily a problem to be solved but a natural outcome of a system that rewards **talent, innovation,** and **entrepreneurship**.

The Redistribution Dimension: A Call for Fairness and Equity

In contrast, the **redistribution dimension** is grounded in the belief that the free market, left unchecked, leads to **economic inequality**, **exploitation**, and **concentrations of wealth** that undermine the well-being of society as a whole. Those who inhabit this dimension argue that **capitalism**, while productive in some ways, tends to create winners and losers in a way that is fundamentally unfair, as it rewards those who already have resources and power while leaving others behind.

In this dimension, the focus is on the need for **economic justice** and **fairness**. Wealth is not viewed solely as the result of individual effort but as something that is often accumulated through structural advantages, **inheritance**, and **systemic inequalities**. The redistribution dimension emphasizes the role of **society** in shaping individual outcomes, recognizing that factors such as education, family background, and access to resources play a significant role in determining one's economic success.

Those in the redistribution dimension argue that the **wealth gap**—the growing divide between the rich and the poor—is a moral failing of the capitalist system. They believe that economic inequality leads to a wide range of social problems, including **poverty**, **crime**, **poor health outcomes**, and **political instability**. In this reality, the concentration of wealth in the hands of a few is seen as not only unjust but also dangerous, as it undermines the **social fabric** and creates a society where opportunities are increasingly limited for the majority.

To address these issues, the redistribution dimension advocates for **government intervention** in the form of **progressive taxation**, **social welfare programs**, and **regulations** that promote economic fairness. The goal is to **redistribute wealth** more equitably, ensuring that everyone has access to basic necessities like healthcare, education, and housing, and that the benefits of economic growth are shared by all, not just the wealthy elite.

This dimension also emphasizes the importance of **collective action** and **solidarity**. Rather than focusing on individual success, the redistribution dimension prioritizes the well-being of the community as a whole. The belief here is that a healthy, functioning society must ensure that everyone has a fair shot at success, and that the economic system should be designed to reduce inequality and uplift the most vulnerable members of society.

The Clash of Economic Realities

The tension between the free market and redistribution dimensions represents a fundamental conflict in how people understand the world. Those in the free market dimension see **capitalism** as a system that rewards **merit** and creates opportunities, while those in the redistribution dimension see it as a system that perpetuates **inequality** and **exploitation**. Each dimension operates with its own set of assumptions about the nature of wealth, power, and opportunity.

In the free market dimension, inequality is often seen as a **necessary** and even beneficial outcome of a competitive system that encourages **innovation** and **productivity**. In this reality, efforts to redistribute wealth through government intervention are viewed with suspicion, as they are believed to create **disincentives** for hard work and stifle economic growth.

Conversely, in the redistribution dimension, inequality is seen as a **symptom** of a system that prioritizes profits over people. In this reality, wealth is often perceived as the result of **structural advantages** and **exploitation** rather than individual effort alone. Advocates for redistribution argue that without intervention, the free market will continue to concentrate wealth in the hands of a few, leaving the majority of people struggling to meet their basic needs.

The clash between these two dimensions is particularly evident in **policy debates** over issues like **taxation, minimum wage laws, universal healthcare,** and **education reform**. In the free market dimension, such policies are often seen as **government overreach**, while in the redistribution dimension, they are viewed as essential steps toward creating a more just and equitable society.

9.2 Capitalism and Socialism as Competing Dimensions

The Capitalist Dimension: Growth, Innovation, and Individual Freedom

For those who inhabit the **capitalist dimension**, the focus is on **economic growth, innovation**, and **individual freedom**. Capitalism is seen as the best economic system because it provides the greatest incentive for individuals to **innovate**, take risks, and **create wealth**. In this dimension, the ability to **own private property** and **pursue profit** is considered essential to personal freedom and economic prosperity.

Capitalism is often praised for its ability to allocate resources efficiently through the mechanisms of **supply and demand**. By allowing individuals and businesses to freely compete in the market, capitalism is believed to foster an environment where the most efficient and innovative solutions to problems naturally rise to the top. In this dimension, **competition** is viewed as a positive force that drives progress and benefits society as a whole.

One of the core values in the capitalist dimension is the belief in the **self-made individual**—the idea that anyone, regardless of their background, can achieve success through **hard work, talent**, and **entrepreneurial spirit**. This narrative of the **American Dream** (or its equivalents in other capitalist societies) is central to the capitalist dimension, as it reinforces the idea that the economic system rewards those who take initiative and work hard.

The capitalist dimension also places a high value on **personal responsibility**. Economic success or failure is often seen as a reflection of an individual's choices, and those who succeed are seen as deserving of their wealth. Conversely, those who struggle economically are often viewed as responsible for their own situation, having made poor choices or failed to take advantage of the opportunities available to them.

The Socialist Dimension: Equality, Collective Well-Being, and Economic Justice

In the **socialist dimension**, the focus shifts from individual wealth and profit to **collective well-being** and **economic justice**. Socialism is rooted in the belief that the capitalist system, while productive in some ways, inevitably leads to **inequality**, **exploitation**, and **concentrations of power** that undermine the well-being of the majority. In this dimension, the goal of the economic system is not to maximize profits but to ensure that everyone has access to the resources they need to live a **dignified** and **fulfilling** life.

The socialist dimension prioritizes **equality** over competition, arguing that the purpose of an economic system should be to **reduce inequality** and promote **fairness**. This dimension often critiques capitalism for allowing a small number of people to accumulate vast wealth while leaving many others in poverty. In response, socialism advocates for a system in which **resources are distributed more equitably**, often through government intervention and collective ownership of key industries.

One of the central tenets of the socialist dimension is the belief in **democratic control of the economy**. Rather than allowing the market to dictate economic outcomes, socialism advocates for a system in which decisions about how resources are allocated are made **collectively** and democratically. This could involve government ownership of certain industries (such as healthcare, education, or utilities) or the establishment of worker-owned cooperatives where employees have a direct say in how their workplace is run.

In the socialist dimension, economic success is not measured by the accumulation of personal wealth but by the overall **well-being** of society. Policies such as **universal healthcare**, **free education**, and **affordable housing** are seen as essential to creating a society where everyone can thrive, regardless of their background or income level. Rather than focusing on individual success, socialism emphasizes the importance of creating a system that works for everyone.

The Conflict Between Capitalism and Socialism

The conflict between **capitalism** and **socialism** represents one of the most profound divides in the **Perceptual Multiverse**. Each dimension operates with its own set of assumptions about the nature of **wealth**, **power**, and **freedom**, and each offers a different vision for how society should be structured.

In the capitalist dimension, **individual freedom** is the highest value, and capitalism is seen as the system that best promotes this freedom. Proponents of capitalism argue that by allowing individuals to freely pursue their own interests, the system creates wealth and prosperity for society as a whole. In this dimension, **competition** is viewed as a natural and beneficial force, and government intervention is seen as a threat to personal liberty.

In contrast, the socialist dimension prioritizes **collective well-being** and **economic justice**. Those in this dimension argue that capitalism leads to **inequality** and **exploitation**, and that the only way to create a fair and just society is through government intervention and collective ownership. In this reality, economic success is measured not by the wealth of individuals but by the health and well-being of the entire society.

The tension between these two dimensions often plays out in **political debates** over issues like **taxation**, **healthcare**, and **social welfare**. In the capitalist dimension, these policies are often viewed as **government overreach** that stifles innovation and reduces individual freedom. In the socialist dimension, however, they are seen as essential steps toward creating a more **equitable** and **just** society.

9.3 How Wealth and Power Shape the Dimensional Divide

The Role of Wealth in Reinforcing Dimensions

Wealth plays a critical role in reinforcing the divide between the free market and redistribution dimensions. In the capitalist dimension, wealth is often seen as a **reward** for hard work and innovation. Those who accumulate wealth are viewed as deserving of their success, and their wealth is seen as a sign of their contribution to society. This dimension celebrates **entrepreneurs**, **business leaders**, and **investors** as the drivers of economic growth and progress.

However, wealth also serves to **entrench** power within this dimension. Those who have accumulated vast wealth often have significant influence over **politics, media,** and **economic policy**. This influence allows the wealthy to shape the economic system in ways that protect their interests and maintain their power. In this way, the capitalist dimension becomes **self-reinforcing**, as the wealthy use their influence to preserve the status quo and resist efforts to redistribute wealth or challenge their dominance.

In the redistribution dimension, wealth is viewed not as a reward for individual effort but as a product of **structural advantages** and **inequalities**. Those who inhabit this dimension argue that the accumulation of wealth by a few comes at the expense of the many, and that the economic system is designed to benefit the wealthy while leaving others behind. In this dimension, wealth is often seen as a **source of power** that corrupts the democratic process and prevents meaningful reform.

Power Dynamics and Economic Inequality

Economic inequality is not just a matter of differences in income or wealth—it is also about **power**. Those who have wealth often have the power to shape the **rules of the game** in ways that benefit themselves and their interests. In the capitalist dimension, this power is seen as a natural outcome of a system that rewards success, while in the socialist dimension, it is viewed as a **corruption** of democracy and a threat to social justice.

In many capitalist societies, the wealthy have significant influence over **political institutions**, often through **campaign contributions, lobbying,** and **control of the media**. This allows them to shape policies that protect their wealth and prevent efforts to redistribute it. For example, wealthy individuals and corporations often lobby against **progressive taxation, labor protections,** and **environmental regulations**, arguing that these policies would harm the economy or stifle innovation.

In the redistribution dimension, this concentration of power is seen as one of the central problems of capitalism. Those in this dimension argue that the wealthy use their power to maintain the status quo, preventing reforms that would benefit the majority of people. In this reality, the concentration of wealth and power in the hands of a few is viewed as fundamentally undemocratic, as it allows a small elite to control the direction of the economy and society.

The Role of Class Consciousness in Shaping the Divide

Another factor that shapes the dimensional divide is **class consciousness**—the awareness of one's position in the economic hierarchy and the recognition of the interests that come with that position. In the capitalist dimension, individuals are often encouraged to see themselves as

potential success stories, with the belief that anyone can achieve wealth and success if they work hard enough. This narrative of upward mobility helps to maintain the capitalist dimension by encouraging individuals to identify with the wealthy rather than with the working class.

In contrast, the socialist dimension often emphasizes the importance of **class solidarity** and the recognition that the interests of the wealthy elite are fundamentally different from those of the working class. In this dimension, individuals are encouraged to see themselves as part of a broader struggle for **economic justice** and to recognize that their interests are aligned with those of other workers, rather than with the wealthy elite.

This difference in **class consciousness** is one of the key factors that perpetuates the divide between the capitalist and socialist dimensions. In the capitalist dimension, the focus on individual success and upward mobility helps to obscure the structural inequalities that keep most people from achieving significant wealth. In the socialist dimension, the emphasis on class solidarity encourages individuals to question the fairness of the economic system and to demand reforms that promote greater equality.

Conclusion: The Ongoing Battle Between Economic Dimensions

The tension between **capitalism** and **socialism**, and between **free markets** and **redistribution**, reflects a broader conflict between two competing dimensions of reality. Each dimension operates with its own set of assumptions about the nature of **wealth**, **power**, and **freedom**, and each offers a different vision for how society should be structured.

In the **capitalist dimension**, wealth is seen as the result of individual effort and merit, and the free market is viewed as the best way to promote innovation and prosperity. In the **socialist dimension**, wealth is often seen as the product of structural inequalities and exploitation, and government intervention is viewed as necessary to create a more just and equitable society.

As we move forward, the **dimensional conflict** between these two realities will continue to shape the way we understand and address issues of **economic inequality**. The future of the economic debate will depend on whether these dimensions can find ways to coexist or whether the tension between them will lead to greater **polarization** and **conflict**.

Ultimately, the question of how to address **economic inequality** is not just about economics—it is about the **moral values** and **beliefs** that underpin our understanding of wealth, power, and justice. By recognizing the competing dimensions at play, we can begin to navigate the complexities of this debate and work toward a more equitable and sustainable future.

Chapter 10: The Political Dimensions of Nationalism vs. Globalism

10.1 National Sovereignty and the Battle for Dimensional Control

The Nationalist Dimension: Protecting Sovereignty and Identity

In the **nationalist dimension**, the most important political value is the **sovereignty** of the nation-state. Nationalists believe that the nation is the primary unit of political organization and that preserving the integrity of the nation is essential to maintaining order, culture, and **self-determination**. This dimension emphasizes the importance of national borders, cultural identity, and the right of each nation to govern itself free from external interference.

In the nationalist dimension, **national sovereignty** is seen as a fundamental right. Nations are viewed as distinct entities with unique histories, cultures, and traditions, and they have the right to pursue their own path without being subject to the dictates of **international organizations** or foreign powers. Nationalists argue that the people of a nation should have the ultimate authority over their own laws, policies, and way of life. Any attempt to undermine or erode national sovereignty—whether through **supranational agreements** or global economic pressures—is seen as a threat to the nation's independence and security.

Nationalism is also closely tied to the idea of **cultural and ethnic identity**. In the nationalist dimension, the nation is often defined by shared language, culture, religion, or ethnicity, and the preservation of this identity is viewed as essential to the survival of the nation itself. As a result, nationalism tends to resist efforts to dilute national identity through **immigration**, **multiculturalism**, or the influence of foreign cultures. Nationalists believe that a strong sense of national identity fosters social cohesion and unity, whereas the loss of this identity leads to fragmentation and societal decline.

The nationalist dimension is particularly resistant to the idea of **global governance**. Institutions like the **United Nations**, the **European Union**, and the **World Trade Organization** are often viewed with suspicion in this reality, as they are perceived as infringing on national sovereignty. Nationalists argue that decisions about **trade, immigration,** and **foreign policy** should be made by the nation's own leaders, not by unelected bureaucrats or international bodies. In this dimension, the priority is on protecting national interests and maintaining control over the nation's destiny.

The Globalist Dimension: Cooperation and Interdependence

In contrast, the **globalist dimension** operates on the belief that the world is increasingly interconnected, and that nations must work together to solve shared problems. In this dimension, **globalization**—the process of integrating economies, cultures, and political systems across borders—is seen as a positive force that fosters **peace, prosperity,** and **cooperation**. Globalists argue that the challenges of the 21st century—such as **climate change, economic inequality,** and **pandemics**—cannot be addressed by nations acting alone. Instead, they require **collective action** and **multilateral cooperation**.

In the globalist dimension, **national sovereignty** is still important, but it is not seen as absolute. Globalists believe that sovereignty must be balanced with the need for **international collaboration** and **mutual responsibility**. While national governments retain the right to make decisions for their own people, they are also part of a global community that requires cooperation on issues that transcend borders. In this dimension, global governance is not about undermining national sovereignty but about creating structures that allow nations to work together on common goals.

Globalization is viewed as a way to bring **nations and people closer together**, fostering cultural exchange, economic growth, and technological innovation. In this reality, the **free flow of goods, people, and ideas** across borders is seen as a positive development that benefits all nations. Globalists argue that **immigration** and **multiculturalism** enrich societies by introducing new perspectives and experiences, and that **international trade** creates opportunities for economic growth and development.

However, the globalist dimension also acknowledges the **challenges** of globalization, such as **economic inequality** and the displacement of jobs due to trade and technological advances. In this reality, these challenges are not reasons to reject globalization but to **reform** it, ensuring that the benefits of global integration are shared more equitably. Globalists advocate for **stronger international institutions** that can regulate global markets, protect human rights, and address environmental concerns. In this dimension, **global cooperation** is seen as essential to building a more just, peaceful, and sustainable world.

The Battle for Dimensional Control

The conflict between nationalism and globalism is not just a disagreement over policy—it is a **battle for dimensional control**. Each dimension offers a different vision of how the world should be organized, and each side believes that its perspective is essential for the future of humanity.

For those in the **nationalist dimension**, globalism represents a threat to national identity and sovereignty. They argue that globalization undermines the ability of nations to control their own destiny, leading to the erosion of **national culture, economic security**, and **political independence**. Nationalists often portray globalism as a force that benefits **elites** at the expense of ordinary citizens, creating a world where **corporations, international institutions**, and **global elites** have more power than democratically elected governments.

In the **globalist dimension**, nationalism is seen as a backward-looking ideology that prevents nations from addressing the complex challenges of the modern world. Globalists argue that retreating into nationalism and **isolationism** will only exacerbate problems like **climate change**, **terrorism**, and **economic inequality**. They believe that **interdependence** is the reality of the 21st century, and that nations must learn to cooperate rather than compete.

This battle for dimensional control is particularly evident in debates over issues like **trade agreements**, **climate change treaties**, and **immigration policy**. Nationalists often resist international agreements that they believe infringe on their country's sovereignty or economic interests, while globalists argue that such agreements are essential for maintaining global stability and addressing shared challenges.

10.2 Globalization: Connecting or Fragmenting Realities?

The Promise of Globalization: Connecting the World

For those in the **globalist dimension**, globalization is seen as a force for **connection** and **progress**. The increasing integration of the world's economies, political systems, and cultures is viewed as a natural and inevitable process that brings people closer together, fosters mutual understanding, and promotes **economic development**.

One of the most important aspects of globalization is the expansion of **international trade**. Globalists argue that trade between nations allows for the more efficient allocation of resources, leading to increased productivity and economic growth. By allowing countries to specialize in the production of goods and services in which they have a comparative advantage, trade fosters **innovation**, reduces costs, and improves the standard of living for people around the world.

Globalization is also seen as a way to **spread democracy** and **human rights**. In this dimension, the integration of nations into a global community creates pressure on authoritarian regimes to adopt democratic reforms and respect human rights. International organizations like the **United Nations** and **nongovernmental organizations** (NGOs) play a key role in promoting these values, and the globalist dimension views the spread of democracy as a key goal of globalization.

Culturally, globalization is viewed as a force that promotes **diversity** and **multiculturalism**. In this dimension, the increased movement of people, goods, and ideas across borders enriches societies by introducing new perspectives and experiences. Globalists argue that multicultural societies are more dynamic, creative, and resilient because they benefit from the contributions of people from different cultural backgrounds.

Technological innovation is another key driver of globalization. Advances in **communication technology**, **transportation**, and the **internet** have made it easier for people to connect with one another, regardless of where they live. In the globalist dimension, technology is seen as a force that breaks down barriers and creates a more interconnected world. The internet, in particular, is viewed as a tool that democratizes access to information and allows people to communicate and collaborate across borders.

The Dark Side of Globalization: Fragmenting Societies

While globalization is celebrated in the globalist dimension as a force for **connection**, those in the nationalist dimension often view it as a source of **fragmentation**. From this perspective, globalization undermines **social cohesion** and creates economic instability, as nations lose control over their own economies and cultures.

One of the most common criticisms of globalization is that it has led to the **outsourcing of jobs** and the decline of traditional industries in many countries. As multinational corporations seek cheaper labor and production costs, they often move manufacturing and other industries to countries with lower wages and fewer regulations. This has led to the **displacement of workers** in industries like manufacturing and agriculture, particularly in developed countries, contributing to **economic inequality** and the decline of the **middle class**.

For those in the nationalist dimension, globalization has also led to the **erosion of national identity**. As countries become more integrated into the global economy and more open to immigration, nationalists argue that the unique cultures, languages, and traditions of individual nations are being diluted. They worry that **multiculturalism** and **immigration** are weakening the social fabric of their countries and creating a sense of alienation among citizens.

Globalization is also seen as a force that benefits **global elites** at the expense of ordinary people. Nationalists argue that while multinational corporations and wealthy individuals have profited from globalization, many working-class people have been left behind. In this dimension, the global economy is seen as rigged in favor of **corporate interests**, and international trade agreements are viewed as tools that benefit large corporations while undermining the livelihoods of workers.

Another major concern in the nationalist dimension is the **loss of political control** to international organizations. Nationalists argue that institutions like the **European Union** and the **World Trade Organization** infringe on the sovereignty of nations by imposing rules and regulations that are not subject to democratic control. In this dimension, globalization is seen as a threat to democracy, as decisions that affect the lives of ordinary citizens are increasingly made by **unelected bureaucrats** in distant international organizations.

The Fragmentation of Realities

While globalization has created unprecedented opportunities for connection, it has also led to the **fragmentation of realities**. The increasing integration of the world's economies, cultures, and political systems has created a situation where individuals and nations experience globalization in vastly different ways, depending on their position in the global economy.

For the **winners of globalization**—typically people in **urban** areas, **multinational corporations**, and individuals with access to education and technology—globalization is a source of opportunity and prosperity. These individuals often thrive in the global economy, benefiting from **new markets, increased mobility**, and **technological innovation**.

However, for the **losers of globalization**—often people in **rural areas, low-skilled workers**, and individuals without access to education and technology—globalization has led to **economic dislocation, cultural alienation**, and a sense of loss. These individuals may feel that their way of life is under threat and that they have little control over the changes taking place around them.

This fragmentation of realities has contributed to the rise of **populist movements** in many countries. These movements, often rooted in the nationalist dimension, draw on the anger and frustration of those who feel left behind by globalization. Populist leaders promise to **restore national sovereignty, protect jobs**, and **defend national identity** against the forces of globalization. In this sense, the fragmentation of realities has fueled the rise of **nationalist politics** and the rejection of globalist ideals.

10.3 How Nations Create Their Own Dimensions of Perception

National Narratives and the Creation of Dimensions

One of the most powerful ways in which nations create their own dimensions of perception is through the use of **national narratives**—stories that shape how a nation's people understand their history, identity, and place in the world. These narratives often emphasize the nation's unique qualities, such as its culture, values, and achievements, and they serve to create a shared sense of identity and purpose.

In the **nationalist dimension**, these narratives are often focused on themes of **sovereignty, self-determination**, and **national pride**. Nationalists emphasize the importance of maintaining control over the nation's destiny, resisting foreign influence, and preserving the nation's cultural identity. They argue that the nation is a distinct and sovereign entity that must be protected from the forces of **globalization** and **external intervention**.

For example, in the United States, the narrative of **American exceptionalism**—the belief that the United States is uniquely suited to lead the world in promoting **freedom, democracy**, and **individual liberty**—has shaped the way many Americans understand their country's role in the world. This narrative reinforces the idea that the United States should protect its sovereignty and pursue its interests independently of international organizations or globalist ideals.

In contrast, in the **globalist dimension**, national narratives often emphasize the nation's role as a part of the global community. Globalists argue that nations are interconnected and that their fates are linked through **trade, diplomacy**, and **shared challenges**. In this dimension, national narratives are focused on themes of **cooperation, multilateralism**, and **interdependence**. Nations are seen as participants in a global system, working together to address common problems like **climate change, terrorism**, and **pandemics**.

For example, in **Germany**, the narrative of **European integration** has been central to the country's post-World War II identity. Germany has positioned itself as a leader in the **European Union**, emphasizing the importance of **cooperation** and **shared governance** in maintaining peace and prosperity on the continent. This narrative reinforces the idea that Germany's future is tied to the success of the European project and that national sovereignty must be balanced with the need for **European unity**.

The Role of Media and Education in Shaping National Dimensions

Media and education play a crucial role in shaping how nations create and sustain their own dimensions of perception. Through **news outlets**, **textbooks**, and **popular culture**, nations propagate the narratives that define their identity and worldview.

In the **nationalist dimension**, media often reinforces themes of **national pride** and **sovereignty**. News outlets may focus on stories that highlight threats to the nation's borders, culture, or economy, while downplaying or criticizing international cooperation or globalist policies. **Patriotic symbols**, **historical achievements**, and **military strength** are often emphasized as central to the nation's identity. In many cases, nationalist media portrays **immigration**, **multiculturalism**, or **international agreements** as dangers to the integrity of the nation.

In the **globalist dimension**, media tends to highlight the importance of **international cooperation** and **shared global challenges**. News outlets may focus on issues like **climate change**, **human rights**, or **global trade**, emphasizing the need for multilateral solutions and collective action. In this dimension, media often portrays **immigration** and **cultural exchange** as positive forces that enrich societies and promote global understanding.

Education systems also play a critical role in shaping national dimensions. In the nationalist dimension, school curricula may emphasize the nation's unique history, culture, and achievements, fostering a sense of national pride and identity. Students may learn about the nation's **heroes**, **wars**, and **struggles for independence**, reinforcing the idea that the nation is a distinct and sovereign entity that must be protected.

In the globalist dimension, education may focus more on **global citizenship** and **multiculturalism**. Students may learn about the importance of **international organizations**, **global trade**, and **human rights**, fostering a sense of interconnection with people from other nations. In this dimension, the emphasis is on preparing students to navigate an increasingly globalized world and to see themselves as part of a broader global community.

Conclusion: The Future of Nationalism vs. Globalism

The tension between **nationalism** and **globalism** represents one of the most significant political divides of the 21st century. Each dimension offers a fundamentally different vision of how the world should be organized and how nations should interact with one another. In the **nationalist dimension**, the focus is on protecting **sovereignty**, **identity**, and **independence**, while in the **globalist dimension**, the emphasis is on promoting **cooperation**, **interdependence**, and **shared responsibility**.

As we move forward, the battle between these dimensions will continue to shape political debates over issues like **immigration**, **trade**, **climate change**, and **international governance**. The future of nationalism vs. globalism will depend on whether these dimensions can find ways to coexist or whether the tension between them will lead to further **polarization** and **conflict**.

Ultimately, the question of how to navigate the competing realities of nationalism and globalism is not just about politics—it is about the **moral values** and **worldviews** that define our understanding of identity, sovereignty, and global responsibility. By recognizing the dimensions at play, we can begin to navigate the complexities of this debate and work toward a more just and interconnected world.

Chapter 11: Digital Dimensions and Information Bubbles

11.1 Social Media as the Gateway to Infinite Dimensions

The Emergence of the Digital Multiverse

The advent of **social media** has radically transformed how individuals interact with the world, allowing people to access, share, and create information at unprecedented speeds. Social media platforms like **Facebook**, **Twitter**, **Instagram**, **YouTube**, and **TikTok** have created virtual spaces where individuals can explore countless topics, join niche communities, and engage in discussions that would have once been geographically or socially limited. Through social media, the boundaries of the physical world have dissolved, giving rise to what can be understood as a **digital multiverse**—a complex and infinite set of dimensions shaped by the thoughts, beliefs, and preferences of individual users.

This **digital multiverse** is characterized by an explosion of information and content, with users exposed to a constant stream of news, opinions, entertainment, and personal updates. However, this seemingly boundless access to information has created both **opportunities** and **challenges**. While social media has democratized access to information and empowered individuals to express themselves, it has also led to the fragmentation of reality, as users are increasingly able to construct their own **personalized digital dimensions**—realities that align with their preferences, beliefs, and biases.

In the digital multiverse, people no longer inhabit a **shared reality** shaped by common sources of information. Instead, they can now choose which dimensions to explore, subscribe to, and participate in. This is both liberating and destabilizing. On one hand, it allows for greater diversity of thought and the amplification of marginalized voices. On the other hand, it creates **information silos** and **echo chambers** where users are exposed only to perspectives that reinforce their existing beliefs, leading to **polarization** and the erosion of common ground.

Social Media as the Architect of New Realities

Social media functions as the primary **gateway** to these digital dimensions. The platforms themselves are not neutral; they are designed to encourage users to engage with content that keeps them on the platform for as long as possible. As a result, the way information is presented on social media plays a critical role in shaping users' perceptions of reality.

Each social media platform has its own unique features that contribute to the creation of **digital dimensions**. For example:

- **Facebook** encourages the formation of communities and interest groups, allowing users to join spaces where they can discuss specific topics and share information with like-minded individuals. These groups often become **echo chambers**, where dissenting views are minimized, and members reinforce each other's beliefs.
- **Twitter** operates as a real-time information hub, where users can follow specific accounts or hashtags to stay updated on the latest trends, news, or political debates. However, the platform's emphasis on **short-form communication** often leads to oversimplification of complex issues and the amplification of emotionally charged content.
- **Instagram** and **TikTok** focus on visual content, creating a space where users can curate their own personal image, follow influencers, and consume entertainment. These platforms play a significant role in shaping users' perceptions of **beauty standards**, **lifestyle aspirations**, and **social norms**.

The architecture of these platforms determines not only what users see but also how they experience and interpret the world around them. Social media acts as a **filter** through which reality is processed and represented, and as users engage with these platforms, they are increasingly drawn into **personalized dimensions** that reflect their interests, beliefs, and desires.

From Shared to Fragmented Realities

Before the rise of social media, traditional forms of media—such as newspapers, television, and radio—helped to create a **shared reality** where citizens received information from common sources. While these traditional media outlets were not without bias, they generally operated under the assumption that they were addressing a **mass audience**. There was an underlying expectation that certain facts, events, and narratives would be understood and accepted by the public at large.

With the rise of social media, this shared reality has fractured into countless **micro-realities**, where users can tailor their information consumption to align with their pre-existing beliefs. In this digital environment, individuals can easily avoid information that challenges their worldview, leading to the creation of **digital bubbles** where they are shielded from dissenting perspectives.

For example, a user who is politically conservative may follow accounts, join groups, and consume content that reinforces conservative beliefs, while a user who is politically liberal may do the same for left-leaning content. As a result, these two users inhabit entirely different **political dimensions**—even though they live in the same physical world, their experiences of reality are shaped by the digital content they consume.

This fragmentation of reality has profound implications for **public discourse** and **democratic society**. When individuals no longer share a common understanding of facts or events, it becomes difficult to engage in meaningful debate or reach consensus on important issues. In this sense, social media has not only opened the door to **infinite dimensions** but also contributed to the **polarization** and **division** of society.

11.2 How Algorithms Create Personalized Realities

The Role of Algorithms in Shaping Digital Experiences

At the heart of social media's ability to create **personalized realities** is the use of **algorithms**—complex mathematical systems designed to filter, sort, and prioritize content based on user behavior. These algorithms are the **invisible architects** of the digital multiverse, determining what users see, what they engage with, and how they navigate their online environments.

Social media algorithms operate by analyzing vast amounts of data about user behavior, including:

- **Likes, shares, and comments**: Each time a user interacts with a post, video, or article, the algorithm learns more about their preferences and adjusts the content they are shown accordingly.

- **Search history and click-throughs**: The content users search for or click on provides additional insights into their interests, which are then used to recommend similar content in the future.
- **Time spent on content**: Algorithms track how long users spend viewing particular types of content, with the assumption that the longer they engage with something, the more interested they are in that subject.
- **Engagement with ads**: User interactions with ads also feed into the algorithm, shaping the types of products, services, or experiences that are recommended to them.

These algorithms are designed to maximize **user engagement**—the amount of time users spend on the platform and how frequently they interact with content. By showing users content that aligns with their preferences and behaviors, algorithms create a **feedback loop** where users are continually exposed to information that reinforces their existing beliefs and interests.

In doing so, algorithms create **personalized realities**—digital dimensions where the content users see is tailored specifically to them. Two users with different preferences and behaviors can have entirely different experiences on the same platform, even if they are following the same news event or topic.

The Echo Chamber Effect

One of the most significant consequences of algorithm-driven personalization is the creation of **echo chambers**—digital spaces where users are exposed primarily to content that confirms their beliefs and biases. In these echo chambers, dissenting views are either minimized or completely absent, allowing users to reinforce their existing worldviews without being challenged by alternative perspectives.

The **echo chamber effect** is particularly pronounced on platforms like **Facebook** and **YouTube**, where users can join communities or subscribe to channels that cater specifically to their interests. Once inside an echo chamber, users are likely to engage with content that aligns with their beliefs, and the algorithm will continue to show them more of the same. This creates a **reinforcement loop** where users are increasingly insulated from opposing viewpoints.

For example, a user who frequently engages with **conspiracy theories** may be shown more conspiracy-related content, regardless of its accuracy or credibility. Over time, this user becomes more deeply entrenched in their belief system, as they are surrounded by content that supports and validates their views. Similarly, a user who primarily consumes content from a particular political ideology will continue to see posts, articles, and videos that reinforce that ideology, leading to a **narrowing** of their perspective.

Echo chambers are not limited to political or ideological beliefs. They can also form around topics like **health**, **lifestyle**, **entertainment**, and **consumer behavior**. In each case, the algorithmic personalization of content contributes to the creation of a **closed-loop reality** where users are shielded from information that might challenge their assumptions or broaden their horizons.

Confirmation Bias and the Algorithmic Feedback Loop

Algorithms also exploit a cognitive tendency known as **confirmation bias**—the human tendency to seek out and interpret information in ways that confirm pre-existing beliefs. Social media platforms, by showing users content that aligns with their preferences, feed directly into this bias, reinforcing users' beliefs and making them less likely to question the validity of the information they encounter.

The **algorithmic feedback loop** works as follows:

1. A user engages with content that reflects their beliefs or interests (e.g., by liking, sharing, or commenting on a post).
2. The algorithm interprets this engagement as a signal that the user is interested in similar content and prioritizes it in their feed.
3. The user continues to engage with this type of content, further reinforcing their preferences.
4. The algorithm responds by showing the user even more content that aligns with their beliefs, narrowing their exposure to alternative viewpoints.

This feedback loop creates a **self-reinforcing cycle** where users are drawn deeper into their personalized realities, making it increasingly difficult for them to encounter or engage with content that challenges their worldview. Over time, this can lead to a **distortion of reality**, as users come to believe that the information they are exposed to on social media represents the full spectrum of truth, rather than just a narrow slice of it.

11.3 The Digital Multiverse: Echo Chambers, Misinformation, and Reality Distortion

The Rise of Misinformation and Disinformation

In the digital multiverse, one of the most pressing challenges is the proliferation of **misinformation** (false or misleading information spread unintentionally) and **disinformation** (deliberately false information spread with malicious intent). Both forms of false information have become pervasive on social media platforms, where the speed and scale of content sharing can quickly turn misinformation into widely accepted "facts."

Misinformation thrives in the **algorithm-driven** environment of social media for several reasons:

- **Viral potential**: Misinformation is often more sensational, emotionally charged, or provocative than factual content, making it more likely to be shared and engaged with. Algorithms prioritize content that generates high levels of engagement, regardless of its accuracy, leading to the widespread dissemination of false information.
- **Confirmation bias**: Users are more likely to believe and share information that aligns with their pre-existing beliefs. Misinformation that confirms a user's worldview is less likely to be scrutinized or fact-checked, especially when it is presented in an echo chamber where opposing viewpoints are scarce.
- **Trust in peer networks**: On social media, users are more likely to trust information shared by their friends, family, or online communities than information from traditional media outlets or expert sources. This creates a fertile ground for the spread of misinformation within closed networks.

Disinformation, on the other hand, is often strategically deployed by actors seeking to **manipulate public opinion**, **sow discord**, or **disrupt democratic processes**. State actors, political operatives, and malicious groups may use social media platforms to spread false or misleading narratives, capitalizing on the fragmented nature of the digital multiverse to reach specific audiences. These efforts are often aimed at **polarizing** society, **undermining trust** in institutions, or advancing particular political agendas.

One of the most prominent examples of disinformation in the digital multiverse is the role of **foreign interference** in elections. During the 2016 U.S. presidential election, Russian operatives used social media platforms to spread disinformation aimed at inflaming political divisions and influencing voter behavior. By targeting specific demographics with tailored content, these actors were able to exploit the **personalized realities** created by social media algorithms, amplifying their impact and distorting public perception of reality.

The Impact of Echo Chambers on Democratic Discourse

The rise of **echo chambers** and the spread of **misinformation** have profound implications for **democratic discourse**. In a democracy, the ability to engage in **open debate**, consider **alternative viewpoints**, and make **informed decisions** is essential. However, when individuals are trapped in echo chambers and exposed primarily to false or misleading information, the foundations of democracy are undermined.

Echo chambers contribute to the **polarization** of society by reinforcing ideological divisions and reducing the likelihood of meaningful dialogue between opposing groups. When users are exposed only to content that aligns with their beliefs, they are less likely to consider alternative perspectives or engage in **constructive debate**. This can lead to an **us vs. them** mentality, where individuals view those who disagree with them not as fellow citizens but as enemies.

Polarization is further exacerbated by the **emotional intensity** of the content that circulates in echo chambers. Content that provokes anger, fear, or outrage is more likely to be shared and engaged with, creating a digital environment where **extreme** or **radical views** are amplified. This emotional intensity makes it more difficult for individuals to approach political or social issues with **nuance** or **critical thinking**, as the focus shifts from rational debate to **emotional validation**.

In the context of **elections** and **public policy**, the distortion of reality caused by echo chambers and misinformation can have serious consequences. When voters base their decisions on **false information** or engage only with content that reinforces their biases, the democratic process is weakened. The result is a society where **truth** becomes subjective, and the ability to reach consensus on important issues becomes increasingly difficult.

The Weaponization of Information in the Digital Multiverse

The digital multiverse has also given rise to the **weaponization of information**, where false or misleading narratives are used as tools of **propaganda, manipulation,** and **psychological warfare**. This phenomenon is particularly evident in the realm of **geopolitics**, where state and non-state actors use social media platforms to advance their strategic objectives.

For example, state-sponsored disinformation campaigns have been used to:

- **Destabilize rival nations** by spreading false narratives that undermine trust in government institutions or create social unrest.
- **Influence elections** by targeting specific voter groups with tailored disinformation designed to sway public opinion or suppress voter turnout.

- **Promote authoritarian regimes** by flooding social media platforms with propaganda that portrays the regime in a positive light while discrediting opposition movements or foreign critics.

In addition to geopolitical actors, **corporate interests** and **special interest groups** have also weaponized information in the digital multiverse. By funding **astroturf campaigns** (fake grassroots movements), **paid influencers**, and **disinformation networks**, these actors can shape public perception of issues like **climate change**, **public health**, or **regulation** in ways that serve their interests.

The weaponization of information poses a serious challenge to the integrity of public discourse and the functioning of democratic societies. When false or misleading narratives are strategically deployed to manipulate public opinion, the result is a **distorted reality** where truth becomes difficult to discern and public trust is eroded.

Conclusion: Navigating the Digital Multiverse

The rise of social media and the proliferation of **algorithms**, **echo chambers**, and **misinformation** have given birth to a **digital multiverse**—a complex and fragmented set of realities where individuals inhabit their own personalized dimensions. While this digital multiverse offers new opportunities for connection, expression, and empowerment, it also presents significant challenges for public discourse, truth, and democracy.

As we move forward, understanding the dynamics of the digital multiverse is essential for navigating the challenges of the 21st century. Efforts to **combat misinformation**, **promote media literacy**, and **foster open dialogue** are critical to preserving the integrity of shared reality in an increasingly fragmented world.

Ultimately, the future of the digital multiverse will depend on our ability to balance the benefits of personalized realities with the need for a **common understanding of truth** and **collective responsibility**. By recognizing the power of **algorithms**, **echo chambers**, and **misinformation**, we can begin to create a digital environment that fosters **informed debate**, **critical thinking**, and **democratic engagement**.

Chapter 12: The Role of Science and Technology in Shaping Dimensions

12.1 Technological Utopias and Dystopias: A Future Reality Split

The Vision of Technological Utopias

As humanity continues to develop ever more advanced technologies, many have envisioned a future where these innovations usher in a **utopian reality**—a world in which technology solves the greatest challenges of our time and creates new possibilities for human flourishing. This **technological utopia** is characterized by the belief that science and technology, when applied responsibly, can lead to the **eradication of disease**, the **elimination of poverty**, and the **enhancement of human potential**.

In this **utopian dimension**, technology is viewed as the ultimate tool for **liberation**. Automation and artificial intelligence (AI) can free people from the burdens of menial labor, allowing them to focus on creative, intellectual, and fulfilling pursuits. Advances in **biotechnology** and **medicine** can extend human life and eliminate suffering from disease. Renewable energy technologies can address the looming threat of **climate change**, enabling humanity to live in harmony with the planet. In this vision, humanity is no longer limited by the constraints of nature but is empowered to shape its own future in ways that were previously unimaginable.

This utopian dimension also extends beyond the physical world. Proponents of technological utopias imagine the possibility of **virtual worlds**, where individuals can transcend the limitations of the body and experience reality in entirely new ways. **Virtual reality (VR)** and **augmented reality (AR)** offer the potential for immersive experiences that blur the line between the digital and physical worlds. In this future, people can explore infinite dimensions of creativity, knowledge, and connection, limited only by their imagination.

In this vision of the future, **AI** plays a central role. AI systems are seen as not only tools for automation but also as partners in human progress. AI could revolutionize **education**, **healthcare**, **governance**, and **scientific research**, accelerating innovation and solving complex problems that have long eluded human effort. In this reality, AI is not feared but embraced as a force for good, guiding humanity toward a future of abundance, equality, and self-actualization.

The Reality of Technological Dystopias

While the vision of technological utopia is appealing, it is accompanied by the **dystopian dimension**, where technology creates as many problems as it solves. In this dystopian reality, advancements in science and technology lead not to liberation but to **surveillance**, **control**, and **inequality**. The technologies that promise to enhance human life are instead used to reinforce existing power structures, deepen inequality, and strip individuals of their autonomy.

One of the most prominent concerns in the dystopian dimension is the rise of **surveillance technology**. Governments and corporations have unprecedented access to personal data, and this data can be used to **monitor**, **manipulate**, and **control** populations. In this dystopian vision, AI systems are not tools for liberation but instruments of oppression. **Facial recognition**, **behavioral tracking**, and **predictive policing** are employed to create a society where every action is monitored, and dissent is suppressed. Privacy becomes a relic of the past, and individuals are reduced to data points in a vast, algorithmic system of control.

In the dystopian dimension, **automation** and **AI** do not lead to a utopia of leisure but to **mass unemployment** and **economic inequality**. As machines take over more jobs, vast swathes of the population are left without meaningful work or economic security. The benefits of technological advancement are concentrated in the hands of a few tech elites, while the majority of people struggle to survive in a world where their labor is no longer valued. This reality is characterized by deepening economic divides, social unrest, and a loss of purpose for many individuals.

Furthermore, the dystopian vision often includes the fear that technology will be used to **manipulate reality** itself. **Deepfake technology**, **AI-generated content**, and **digital misinformation** can blur the line between truth and fiction, making it increasingly difficult to discern what is real. In this dystopian future, individuals are bombarded with conflicting narratives, and reality becomes a contested space, shaped by those with the power to control information.

The Tension Between Utopia and Dystopia

The tension between these **utopian** and **dystopian** dimensions reflects the ambivalence that many people feel about the rapid pace of technological change. On one hand, there is a deep belief in the **potential** of technology to improve human life and create a better world. On the other hand, there is an equally strong fear that technology could lead to **unintended consequences**, reinforcing existing inequalities and creating new forms of control and alienation.

This tension is particularly evident in debates over emerging technologies like **artificial intelligence**, **genetic engineering**, and **biotechnology**. These technologies hold the promise of **revolutionizing** medicine, agriculture, and industry, but they also raise profound ethical questions about **who controls these technologies** and how they will be used. Will the benefits of these innovations be shared equitably, or will they exacerbate existing divides between the rich and the poor? Will technology enhance individual freedom, or will it become a tool for surveillance and control?

The future of these **competing dimensions** will depend largely on how society chooses to navigate the ethical challenges posed by technological advancement. As we explore these questions, it becomes clear that the line between **utopia** and **dystopia** is often thin, and the outcome of the technological revolution will be shaped by the decisions we make today.

12.2 Artificial Intelligence: The Catalyst for a New Dimension

AI as a Revolutionary Force

Artificial intelligence (AI) is often seen as the most transformative technology of the 21st century—a catalyst that could reshape every aspect of society and create a new dimension of reality. In this new dimension, AI serves as the foundation for **economic**, **social**, and **scientific innovation**, accelerating progress at a pace previously unimaginable.

In the realm of **economics**, AI has the potential to revolutionize industries by automating processes, optimizing production, and driving innovation. AI systems can analyze vast amounts of data, uncovering patterns and insights that humans might overlook. This capability can lead to the creation of new products, services, and business models that transform industries ranging from **finance** to **healthcare**. AI-driven systems can also make supply chains more efficient, reducing waste and improving sustainability.

In **scientific research**, AI is already proving to be a game-changer. AI systems can analyze complex data sets and generate new hypotheses, accelerating the pace of discovery in fields such as **genomics**, **drug development**, and **materials science**. In this new dimension, AI acts as a collaborator with human scientists, helping to solve some of the most pressing challenges in medicine, climate science, and engineering. For example, AI-driven models have been used to predict protein folding, a breakthrough that has the potential to revolutionize drug development and disease treatment.

In the social realm, AI could enhance human life by improving access to **education**, **healthcare**, and **government services**. AI-driven educational platforms can personalize learning experiences, tailoring content to the needs and abilities of individual students. In healthcare, AI systems can assist doctors in diagnosing diseases, recommending treatments, and managing patient care. In government, AI can streamline bureaucratic processes, making public services more efficient and accessible to all citizens.

The Birth of a New Reality: Human-AI Collaboration

As AI systems become more advanced, the line between **human intelligence** and **machine intelligence** begins to blur. In this new reality, AI is not just a tool used by humans; it is a collaborator, capable of working alongside people to solve complex problems and create new possibilities.

One of the most exciting aspects of this new dimension is the potential for **human-AI collaboration**. AI systems can process vast amounts of information, identify patterns, and make recommendations, while humans bring **creativity**, **empathy**, and **moral judgment** to the table. Together, humans and AI can achieve things that neither could accomplish alone.

For example, in the field of **medicine**, AI systems can analyze medical records, scan images, and recommend treatment plans based on data from millions of patients. However, doctors still play a crucial role in interpreting this data, making ethical decisions, and providing care with empathy and understanding. In this new reality, the partnership between humans and AI allows for **personalized medicine** that is tailored to the unique needs of each patient.

Similarly, in **education**, AI-driven platforms can deliver personalized lessons that adapt to each student's learning style and pace. Teachers, however, remain essential for providing **emotional support**, fostering **critical thinking**, and creating a sense of community in the classroom. In this new dimension, the combination of AI's ability to deliver information at scale and human teachers' capacity for connection and mentorship creates a more **holistic educational experience**.

In this new reality, AI is not seen as a replacement for human labor but as a **partner** that augments human capabilities. This collaboration opens up new possibilities for **creativity**, **problem-solving**, and **innovation**, allowing humanity to tackle challenges that were once considered insurmountable.

The Ethical Challenges of AI

While the potential of AI is enormous, it also raises significant ethical questions about how this new dimension will be shaped. As AI systems become more integrated into every aspect of society, concerns about **bias, transparency**, and **accountability** have come to the forefront.

One of the most pressing ethical challenges is the issue of **bias** in AI systems. AI algorithms are trained on data sets that reflect the biases present in the real world, and as a result, they can unintentionally reinforce or amplify these biases. For example, AI systems used in hiring processes have been found to discriminate against certain groups based on gender, race, or socioeconomic status. In the criminal justice system, AI-driven predictive policing algorithms have been criticized for disproportionately targeting minority communities.

To address these ethical concerns, it is essential to develop AI systems that are **transparent, fair**, and **accountable**. This requires not only technical solutions—such as improving the quality of data used to train AI models—but also broader societal efforts to ensure that AI is used in ways that promote **equity** and **justice**. As AI continues to shape this new dimension of reality, the decisions we make about how to design, deploy, and regulate these systems will have profound implications for the future of society.

Another ethical challenge is the question of **accountability** in AI-driven systems. As AI takes on more responsibilities in areas like healthcare, transportation, and law enforcement, it becomes increasingly important to determine who is responsible when things go wrong. If an AI system makes a decision that leads to harm—such as a misdiagnosis in a medical setting or an error in a self-driving car—who should be held accountable? Is it the developers who created the system, the organizations that implemented it, or the AI system itself? These questions are becoming more urgent as AI systems take on more critical roles in society.

12.3 The Ethics of Dimensional Expansion Through Technology

Expanding Human Capabilities Through Technology

One of the most exciting aspects of technological advancement is its potential to **expand human capabilities** and open up new dimensions of experience. From **genetic engineering** to **cybernetics**, technologies that enhance or augment human abilities could fundamentally change what it means to be human, allowing individuals to transcend the limitations of biology and unlock new potential.

Biotechnology offers the promise of extending human life, curing genetic diseases, and even enhancing cognitive abilities. Through techniques like **CRISPR gene editing**, scientists are beginning to unlock the ability to manipulate DNA, potentially curing hereditary diseases like **cystic fibrosis** or **sickle cell anemia**. In the future, these technologies could be used not only to treat diseases but also to enhance human traits, such as **intelligence**, **physical strength**, or **longevity**.

Similarly, advances in **neurotechnology** could allow for the direct interface between the human brain and machines. Technologies like **brain-computer interfaces (BCIs)** are already being developed to help people with disabilities control devices with their thoughts, and in the future, these interfaces could enable seamless communication between humans and machines. This could open up entirely new dimensions of experience, allowing individuals to access information, control technology, and even communicate telepathically.

However, these possibilities raise profound ethical questions about the consequences of **enhancing human capabilities**. Should technologies that enhance cognitive or physical abilities be available to everyone, or will they be accessible only to the wealthy, exacerbating existing inequalities? What are the potential risks of altering human biology in ways that could have unforeseen consequences? And how should society regulate the development and use of these technologies to ensure they are used ethically?

The Moral Implications of Transcending Physical Reality

As technology continues to advance, the possibility of **transcending physical reality** through **virtual environments** and **digital consciousness** becomes more plausible. The development of **virtual reality (VR)** and **augmented reality (AR)** technologies has already created new dimensions of experience, allowing individuals to explore virtual worlds that are limited only by their imagination. In these digital dimensions, individuals can create new identities, interact with others in novel ways, and experience realities that are fundamentally different from the physical world.

In the future, the concept of **uploading consciousness** into a digital environment—sometimes referred to as **mind uploading** or **digital immortality**—could enable individuals to live indefinitely in a virtual dimension. This would represent a radical expansion of human experience, as individuals could transcend the limitations of the physical body and explore entirely new forms of existence.

However, the ethical implications of such technologies are profound. What does it mean to be **human** in a world where consciousness can exist outside the body? How would these technologies affect our understanding of **life**, **death**, and **identity**? And what are the potential risks of creating a reality where individuals can escape into virtual worlds, leaving behind the challenges and responsibilities of the physical world?

The development of **digital dimensions** also raises questions about the potential for **exploitation** and **inequality**. In a world where individuals can escape into virtual realities, there is a risk that these technologies could be used to distract or pacify populations, diverting attention from pressing social, economic, or political issues. Additionally, the creation of **virtual economies** within these digital dimensions could lead to new forms of inequality, where those who control the digital infrastructure hold immense power over those who inhabit these virtual worlds.

The Responsibility of Dimensional Creators

As we expand into new dimensions through technology, the responsibility of those who design, build, and regulate these technologies becomes increasingly important. **Technologists**, **scientists**, **engineers**, and **policymakers** must grapple with the ethical implications of their work and consider how the technologies they create will shape society.

One of the central ethical questions in this new dimension is the issue of **equity**. As technologies like AI, genetic engineering, and virtual reality continue to develop, there is a risk that these innovations will primarily benefit those who are already privileged, leaving behind those who lack access to education, resources, or opportunities. Ensuring that the benefits of technological advancement are **shared equitably** is essential for creating a future where new dimensions of reality are accessible to all.

Another important ethical consideration is the need for **transparency** and **accountability** in the development of new technologies. As we move into uncharted territory, it is crucial that the processes and decisions behind technological innovations are open to scrutiny. This includes ensuring that **AI systems** are designed and deployed in ways that are fair and transparent, that genetic engineering is conducted ethically and responsibly, and that virtual environments are governed by principles that protect individual autonomy and freedom.

Conclusion: The Future of Dimensions Shaped by Science and Technology

As we explore the intersection of science, technology, and dimensional realities, it becomes clear that the future of human experience will be profoundly shaped by the choices we make today. **Technological utopias** and **dystopias** represent competing visions of what is possible, and the path we take will depend on how we navigate the ethical challenges of technological advancement.

The rise of **artificial intelligence** has the potential to catalyze a new dimension of reality—one where human-AI collaboration unlocks unprecedented possibilities for creativity, innovation, and problem-solving. However, the development of AI also raises important questions about bias, accountability, and the potential for exploitation.

As technology continues to expand human capabilities and create new dimensions of experience, we must grapple with the profound **ethical implications** of dimensional expansion. Ensuring that these technologies are used in ways that promote equity, justice, and human flourishing will be essential for shaping a future where science and technology enhance, rather than undermine, the human experience.

Chapter 13: Crossing the Dimensional Boundaries

13.1 How Do We Shift from One Dimension to Another?

Dimensions of Thought and Belief: A Multiverse of Perception

In the **Perceptual Multiverse**, dimensions are not just theoretical constructs but reflect the myriad ways in which individuals perceive reality. These dimensions are shaped by **thought patterns**, **belief systems**, **cultural narratives**, and **personal experiences**. When we talk about **crossing dimensional boundaries**, we are referring to the process of shifting from one set of perceptions, values, or beliefs to another, often in response to new information, experiences, or internal conflicts.

Each dimension can be thought of as a self-contained reality in which the inhabitants share common **cultural assumptions**, **social norms**, and **ideological frameworks**. Within any dimension, certain **truths** or **beliefs** are considered fundamental and unchallenged, shaping how individuals interpret the world around them. However, the **boundaries between**

dimensions are not fixed; individuals can, and often do, move between dimensions as their **beliefs** or **perceptions** shift over time.

Triggers for Dimensional Shifts

There are several ways in which individuals might shift from one dimension to another:

1. **Exposure to New Information or Perspectives**: One of the most common ways to shift between dimensions is through the introduction of **new information** or perspectives that challenge an individual's existing beliefs. This can happen through **education**, **travel**, **personal experiences**, or even exposure to new ideas through **books**, **media**, or **conversations**. For example, someone who has lived their entire life within a specific political or religious belief system might begin to question that system after encountering information that contradicts what they have been taught.
   - **Example**: A person who has spent their life in a dimension where climate change is denied may begin to shift into a new dimension after watching a compelling documentary that presents scientific evidence for climate change. This new information challenges their previous understanding, and over time, they may begin to see the world through the lens of environmental responsibility.
2. **Personal Crises or Transformative Experiences**: Major life events—such as **illness**, **loss**, **trauma**, or **spiritual experiences**—can also serve as catalysts for dimensional shifts. These events force individuals to re-evaluate their existing beliefs, often leading them to adopt new perspectives or worldviews.
   - **Example**: A person who has lived in a dimension that prioritizes material success may undergo a shift after experiencing a health crisis, leading them to adopt a new dimension that values wellness, spirituality, and inner peace over financial achievement.
3. **Gradual Evolution of Beliefs**: Not all dimensional shifts happen suddenly. For many people, shifts occur as part of a **gradual evolution** of beliefs and values over time. As individuals grow older, experience new things, or encounter different viewpoints, their beliefs may naturally evolve, leading them to cross into new dimensions of perception.
   - **Example**: A person who begins their career with a strong belief in individualism and the free market may, over time, shift toward a more communal or socialist worldview as they witness the impact of economic inequality on their community. This shift may be subtle, happening over years, but it represents a significant movement from one dimension to another.
4. **Social Influence and Group Dynamics**: Dimensional shifts can also be driven by **social influences** or **group dynamics**. People are often influenced by the beliefs and behaviors of those around them, and belonging to a particular social group can reinforce certain dimensions of thought while making it more difficult to adopt others. However, when individuals switch social circles, join new communities, or are exposed to different cultures, they may begin to adopt the beliefs and perspectives of those new groups, leading to a shift in dimensions.
   - **Example**: A student who grows up in a conservative household may shift toward more progressive views after attending college and joining a community that emphasizes diversity and inclusion. The influence of their peers, professors, and new experiences in this environment can gradually lead them to cross the boundaries into a different dimension of thought.

Barriers to Dimensional Shifts

While crossing dimensional boundaries is possible, it is not always easy. Several **barriers** can prevent individuals from shifting into new dimensions, even when they are exposed to new information or experiences:

1. **Cognitive Inertia**: Human beings have a natural tendency to stick with what they know. This **cognitive inertia** makes it difficult to change deeply held beliefs, especially if those beliefs have been reinforced over time by family, culture, or personal experience.
2. **Fear of the Unknown**: Shifting into a new dimension often involves **uncertainty**. People may resist crossing dimensional boundaries because they fear losing their sense of identity or the security that comes with holding onto familiar beliefs. The prospect of adopting a new worldview can be unsettling, especially if it requires letting go of deeply ingrained values or assumptions.
3. **Social and Cultural Pressures**: Many individuals are reluctant to shift dimensions because of the **social consequences**. Belonging to a community, whether it is a religious group, political party, or cultural identity, often comes with certain expectations about what beliefs are acceptable. Crossing into a new dimension may mean facing **social alienation**, **criticism**, or even **rejection** from one's peers or family members.
4. **Cognitive Dissonance**: When individuals are confronted with information that contradicts their existing beliefs, they often experience **cognitive dissonance**—a psychological discomfort that arises from holding conflicting ideas. In many cases, people resolve cognitive dissonance by rejecting or dismissing the new information, rather than allowing it to trigger a dimensional shift. This defense mechanism helps individuals maintain consistency in their beliefs but can also prevent them from exploring new dimensions of reality.

13.2 Cognitive Dissonance as a Symptom of Dimensional Conflict

Understanding Cognitive Dissonance

Cognitive dissonance is a key concept in understanding how individuals respond to the **conflict** between competing dimensions of reality. When a person holds two or more contradictory beliefs or values, or when their behavior conflicts with their beliefs, they experience cognitive dissonance. This internal conflict creates psychological discomfort, which individuals are motivated to resolve.

In the context of the Perceptual Multiverse, cognitive dissonance often arises when an individual encounters **evidence**, **ideas**, or **experiences** that contradict the beliefs or values of the dimension they currently inhabit. This dissonance can act as a **catalyst for change**, encouraging the individual to shift into a new dimension, but it can also lead to **defensive reactions** that reinforce their existing beliefs.

Dissonance and the Reinforcement of Dimensions

For many people, cognitive dissonance does not lead to a dimensional shift. Instead, they resolve the dissonance by rejecting or minimizing the **contradictory information** in order to preserve their existing beliefs. This process is known as **cognitive bias**, and it includes strategies such as:

1. **Confirmation Bias**: The tendency to seek out information that confirms one's existing beliefs while ignoring or dismissing information that contradicts them. In the context of dimensional conflict, confirmation bias allows individuals to reinforce the boundaries of their current dimension by filtering out challenging perspectives.
2. **Selective Exposure**: This occurs when individuals consciously or unconsciously expose themselves only to media, ideas, or communities that align with their beliefs. By limiting their exposure to opposing viewpoints, individuals can avoid cognitive dissonance and remain securely within the boundaries of their dimension.
3. **Rationalization**: When faced with conflicting information, individuals may attempt to rationalize away the contradictions by finding explanations that allow them to maintain their existing beliefs. This can involve reinterpreting facts, blaming external factors, or using **mental gymnastics** to reconcile inconsistencies.
- **Example**: A person who is deeply invested in a political ideology may encounter evidence that contradicts one of the central tenets of that ideology. Rather than shifting to a new dimension, they may rationalize the inconsistency by attributing the evidence to **media bias** or dismissing it as an outlier, thus preserving their current belief system.

Dissonance and Dimensional Shifts

While cognitive dissonance often reinforces existing dimensions, it can also serve as a trigger for dimensional shifts. When the dissonance becomes too great to ignore or resolve, individuals may be forced to re-evaluate their beliefs and consider new perspectives. In these cases, cognitive dissonance acts as a **crack in the foundation** of their current dimension, opening the possibility for change.

There are several factors that can determine whether cognitive dissonance leads to a dimensional shift:

1. **The Strength of the Dissonance**: The greater the contradiction between new information and existing beliefs, the more intense the cognitive dissonance. When dissonance reaches a critical threshold, it can become impossible to ignore, leading to a re-evaluation of one's beliefs.
2. **Openness to New Ideas**: Individuals who are naturally curious or open-minded are more likely to allow cognitive dissonance to lead to a dimensional shift. These individuals are willing to entertain new ideas, even if they challenge their existing worldview.
3. **Social Support**: Dimensional shifts are often easier when individuals have the support of others who share their new beliefs. Conversely, the absence of support can make it more difficult to cross dimensional boundaries, as individuals may fear **social isolation** or **rejection**.
4. **Personal Reflection**: Engaging in deep personal reflection—whether through **meditation**, **journaling**, or **discussion**—can help individuals work through cognitive dissonance and come to terms with new perspectives. This reflective process often precedes a shift from one dimension to another, as it allows individuals to process conflicting ideas in a thoughtful and deliberate way.
- **Example**: A religious individual who experiences cognitive dissonance after learning about scientific evidence for evolution may spend months or years reflecting on the relationship between their faith and science. Over time, they may shift into a new dimension where they reconcile their spiritual beliefs with scientific knowledge.

13.3 Can We Find Common Ground Between Parallel Realities?

The Fragmentation of Reality

One of the most pressing challenges in the modern world is the **fragmentation of reality** into parallel dimensions of thought and belief. This fragmentation is evident in the way that individuals and groups live in **separate informational bubbles**, consume different media, and adopt opposing narratives about political, social, and cultural issues.

In some cases, these parallel realities are so divergent that finding **common ground** seems impossible. People from different dimensions may not only disagree about solutions to problems but also about the **nature of the problems** themselves. For example, two individuals may inhabit completely different dimensions when it comes to issues like **climate change**, **immigration**, or **vaccines**—one dimension may view climate change as an existential threat, while another denies its existence altogether.

Despite this fragmentation, the question remains: Is it possible to find common ground between these parallel realities, and if so, how?

The Role of Dialogue in Bridging Dimensions

One of the most promising ways to bridge the gap between parallel dimensions is through **open dialogue**. Meaningful communication between individuals from different dimensions of thought can help to break down barriers, foster understanding, and create opportunities for **shared perspectives**.

However, dialogue is not always easy, especially when individuals are deeply entrenched in their beliefs. For dialogue to be effective, several key conditions must be met:

1. **Empathy**: Individuals must approach the conversation with **empathy** and a willingness to understand the other person's perspective. This requires setting aside preconceived notions and being open to hearing why the other person holds their beliefs.
2. **Active Listening**: Effective dialogue requires **active listening**—paying attention to what the other person is saying without immediately formulating a rebuttal. By listening carefully, individuals can gain a better understanding of the reasoning and experiences that have shaped the other person's dimension.
3. **Common Values**: While individuals from different dimensions may disagree on specific issues, they often share common values—such as a desire for **justice**, **security**, or **well-being**. Identifying and emphasizing these shared values can create a foundation for constructive dialogue and collaboration.
4. **Respectful Disagreement**: Not all dialogue will lead to agreement, but respectful disagreement is still valuable. By engaging in **civil discourse** and acknowledging the legitimacy of different perspectives, individuals can build **mutual respect**, even when they do not see eye to eye.
5. **Framing the Debate**: Reframing discussions around common goals or outcomes can help bridge the gap between dimensions. For example, instead of debating whether **climate change** is real, individuals might focus on shared concerns about **pollution**, **energy security**, or **economic development**.
- **Example**: In a discussion about immigration, instead of focusing on divisive rhetoric, participants might agree that **border security** and **humanitarianism** are both important.

By reframing the debate in terms of shared values, individuals from different dimensions can find areas of overlap.

Building Shared Realities

While dialogue is essential, it may not be sufficient to fully bridge the divide between parallel dimensions. To truly find **common ground**, societies must work toward building **shared realities** that are grounded in **evidence**, **facts**, and **mutual understanding**.

One approach to building shared realities is through the promotion of **critical thinking** and **media literacy**. In a world where misinformation and disinformation are rampant, helping individuals develop the skills to critically evaluate information is crucial. By encouraging people to question their sources, verify facts, and engage with a variety of perspectives, societies can begin to rebuild trust in a shared reality.

Another important factor is the role of **education**. Schools and universities have a responsibility to foster an environment where students are exposed to diverse ideas and taught how to engage in respectful debate. **Interdisciplinary education**, which combines science, history, philosophy, and the arts, can help students develop a well-rounded understanding of the world and resist the pull of echo chambers.

Finally, the media and technology companies must play a role in promoting **balanced discourse** and preventing the spread of **polarizing content**. Algorithms that prioritize sensational or extreme viewpoints over balanced reporting contribute to the fragmentation of reality. By promoting **responsible journalism** and **transparent algorithms**, these companies can help create a digital environment where individuals are more likely to encounter multiple perspectives.

The Challenges of Bridging Parallel Realities

Despite these efforts, the task of bridging parallel realities is not without challenges. The **political**, **economic**, and **social forces** that contribute to the fragmentation of reality are deeply entrenched, and many individuals may resist efforts to find common ground. Additionally, in some cases, the differences between dimensions are so profound that no amount of dialogue or education can fully reconcile them.

However, even in these cases, the goal should not be to **erase** differences but to create a society where individuals from different dimensions can **coexist** peacefully. By fostering an environment of **tolerance**, **respect**, and **understanding**, societies can ensure that even in a world of parallel realities, there is room for diverse perspectives and meaningful collaboration.

Conclusion: Crossing the Dimensional Boundaries

Crossing the boundaries between dimensions of thought and belief is both a **personal** and **societal** challenge. For individuals, it involves the willingness to engage with new ideas, confront cognitive dissonance, and question deeply held assumptions. For societies, it requires creating spaces for dialogue, promoting critical thinking, and building shared realities that are grounded in evidence and mutual understanding.

While the task of bridging parallel realities is daunting, it is not impossible. By recognizing the ways in which individuals shift between dimensions, addressing the role of cognitive dissonance, and fostering open dialogue, we can begin to create a world where individuals from different dimensions can find common ground and work together toward a shared future.

Chapter 14: Extra Dimensions Beyond the Flashpoints

14.1 The Hidden Dimensions Beyond Physical Perception

The Concept of Dimensions Beyond Our Immediate Senses

The notion that reality is far more expansive than what we perceive through our five physical senses is a central theme in both scientific inquiry and spiritual traditions. While we experience the world through sight, sound, touch, taste, and smell, these faculties only provide access to a narrow **bandwidth** of the **full spectrum of reality**. Throughout history, mystics, shamans, and thinkers have posited that there are hidden dimensions, realities that exist beyond the **veil of perception**, accessible only through **expanded consciousness**, spiritual practice, or scientific exploration.

The idea of **extra dimensions**—realities that exist alongside, above, or beyond our own—has found its way into both **philosophical thought** and **scientific theories**. The ancient Greeks, for example, spoke of **higher realms of being**, while Eastern philosophies like **Hinduism** and **Buddhism** describe **layers of existence** beyond the material world. In recent times, **science** has echoed these ideas, particularly in **theoretical physics**, where discussions about additional dimensions have become central to understanding the fabric of the universe.

One of the core assumptions behind the existence of extra dimensions is that our **immediate senses** are limited in their scope. Humans are biologically programmed to perceive a specific range of stimuli—visible light, audible sound, and tactile sensations that allow us to interact with the world. But this is only a fraction of the larger **energy spectrum** that exists. For instance, ultraviolet light and infrared radiation are invisible to us, even though they are integral parts of the **electromagnetic spectrum**. Similarly, other dimensions may be present, but we simply lack the biological apparatus or the technology to **perceive them directly**.

The question then becomes: How can we transcend these sensory limitations to perceive and interact with the **full scope of reality**? And if we could, what would we find?

The Limitations of the Human Mind in Perceiving the Full Scope of Reality

The limitations of human perception are not merely biological but also **cognitive**. The **mind** processes a staggering amount of information every second, but much of it is filtered through **cognitive shortcuts** that allow us to make sense of the world without being overwhelmed. As a result, much of the **raw data** that could potentially point to other dimensions or realities is **discarded** or interpreted in ways that align with our preconceived notions of reality.

Neuroscientific studies show that the brain operates using **predictive coding**, meaning that it creates models of the world based on past experiences and expectations. These models shape what we perceive, filtering out anything that doesn't fit into the familiar patterns. This cognitive mechanism is helpful for navigating daily life, but it also limits our ability to experience **unfamiliar dimensions** of reality. As a result, the **mind** can become a gatekeeper, locking us into specific perceptual and conceptual frameworks that align with our current understanding of the world, while excluding information that points to hidden or alternate dimensions.

For example, many spiritual traditions suggest that **altered states of consciousness**—such as **meditation**, **dreams**, or **psychedelic experiences**—can momentarily bypass the brain's filters, allowing us to perceive **realities** that are normally hidden. These experiences suggest that **consciousness** itself may be capable of **traversing dimensions** that are inaccessible to ordinary waking perception. In this way, the mind serves as both the **gateway** and the **limiting factor** in our exploration of extra dimensions.

Scientific Theories on Dimensions We Cannot Observe

Modern science, particularly in the realm of **theoretical physics**, has begun to explore the possibility of hidden dimensions through complex mathematical models and thought experiments. The most prominent of these is **string theory**, which proposes that the fundamental building blocks of the universe are not **particles** but **vibrating strings**. These strings, it is theorized, exist in **multiple dimensions**, some of which are **curled up** or **compactified** in such a way that they are imperceptible to human senses or current technology.

String theory, along with its extension known as **M-theory**, suggests that the universe may have as many as **10 or 11 dimensions**. In our everyday experience, we are familiar with three spatial dimensions—**length**, **width**, and **height**—plus the dimension of **time**. However, according to string theory, the additional dimensions are hidden because they are folded up at scales so small that they are undetectable by current instruments.

This idea introduces a fascinating possibility: that **other dimensions** could be influencing our reality in ways that we cannot directly perceive. These extra dimensions may hold the key to solving some of the most profound mysteries in physics, such as the nature of **dark matter**, the unification of forces, or the **origin of the universe**. If these dimensions exist, they are as real as the dimensions we experience, yet they remain invisible, operating beyond the reach of our current technological and cognitive capabilities.

14.2 Vibrational Dimensions: How Frequency Shapes Reality

The Idea That Different Dimensions Vibrate at Different Frequencies

The concept of **vibration** as a fundamental aspect of reality has long been a theme in both **spiritual traditions** and **modern science**. In recent years, the idea of **vibrational dimensions** has gained traction, particularly in **metaphysical** and **energy-based philosophies**. According to this view, everything in the universe—whether physical matter, light, sound, or thought—vibrates at specific frequencies. The notion that dimensions are organized by **vibrational frequencies** suggests that reality itself is multi-layered, with different levels of existence or dimensions vibrating at distinct rates.

In physics, the concept of **frequency** is already well understood. Sound, light, and electromagnetic waves are all measured by their frequencies. The higher the frequency, the more energy the wave contains. Applying this concept to dimensions, some theorize that **higher-dimensional planes** may exist at **higher vibrational frequencies**, which is why they are imperceptible to us in our current state. According to this theory, humans naturally resonate within a certain range of frequencies—those that correspond to the **physical dimension**—but other dimensions exist beyond this range.

For example, in the **spiritual dimensions** often described by **mystics** and **energy healers**, beings and entities are said to vibrate at frequencies higher than our physical dimension. These beings, whether they are referred to as **angels**, **spirit guides**, or **higher selves**, exist in dimensions that are invisible to our ordinary senses because they operate at a different vibrational level. Similarly, in many Eastern philosophies, it is believed that **enlightenment** or **spiritual awakening** involves raising one's vibrational frequency to align with higher planes of existence.

How Altering Our Personal Vibrational Frequency Can Allow Us to Access New Realities

One of the key ideas in vibrational theory is that **altering** our personal vibrational frequency can allow us to access **new realities**. This idea suggests that our **consciousness** and **energy** are not fixed, but are instead fluid and adaptable. By shifting our vibration—whether through **meditation**, **energy work**, or other practices—we can attune ourselves to different levels of reality.

In many spiritual practices, the concept of **energy centers** or **chakras** plays an important role in understanding how vibrational shifts occur. The chakras are believed to be centers of energy in the body, each corresponding to different aspects of physical, emotional, and spiritual life. When these chakras are balanced and vibrating at their optimal frequency, individuals experience heightened awareness, greater spiritual connection, and access to higher dimensions of existence. Techniques such as **breathwork**, **sound healing**, and **visualization** are often used to raise the vibrational frequency of these energy centers, allowing individuals to enter into expanded states of consciousness.

From a metaphysical perspective, it is believed that **higher-vibrational states** open the door to alternate realities and dimensions. For example, during deep meditation or moments of **spiritual clarity**, individuals may experience visions, insights, or sensations that transcend ordinary reality. These experiences are often interpreted as **glimpses** into higher-dimensional realms, where the laws of physical matter do not apply, and where **thought** and **intention** take on a more tangible form.

- **Example**: Many spiritual traditions describe the experience of **astral projection**, where an individual's consciousness separates from their physical body and travels to other realms. This practice is often associated with raising one's vibrational frequency to match the energy of the **astral plane**, which exists beyond the physical dimension.

Techniques for Tuning into Different Vibrational States

There are numerous techniques for **tuning into** different vibrational states, many of which have been practiced for centuries in spiritual traditions around the world. These techniques aim to raise the practitioner's vibrational frequency, allowing them to access higher dimensions of reality.

1. **Sound Therapy**: One of the most powerful ways to alter one's vibrational frequency is through **sound therapy**. Sound, being a form of vibration, has the ability to resonate with the energy centers in the body, bringing them into alignment with higher frequencies. Techniques such as using **Tibetan singing bowls**, **binaural beats**, and **chanting** are all ways to entrain the mind and body to higher vibrational states.
    - **Tibetan Singing Bowls**: These metal bowls are struck or circled with a mallet to create resonant sounds that correspond to specific frequencies. Practitioners believe that these frequencies can harmonize the body's energy centers and elevate one's vibrational state.

- **Binaural Beats**: By playing two slightly different frequencies in each ear, binaural beats create the illusion of a third frequency that the brain entrains to. This can induce altered states of consciousness, promoting deep relaxation, meditation, or even access to higher dimensions.
2. **Meditation and Breathwork**: **Meditation** is one of the most effective tools for raising one's vibrational frequency. By quieting the mind and focusing inward, individuals can connect with their **higher self** and align their energy with higher dimensions. Specific forms of meditation, such as **mantra meditation** or **breath-focused meditation**, are often used to elevate consciousness.
    - **Breathwork**: Conscious breath control, such as **pranayama** in yoga, can alter the body's vibrational frequency by increasing **oxygen flow**, balancing energy, and clearing emotional blockages. Breathwork techniques can help individuals enter states of heightened awareness and vibrational attunement.
3. **Energy Healing**: Techniques such as **Reiki** and **qigong** involve the manipulation of **subtle energy fields** to bring the body into harmony with higher frequencies. These practices are often used to clear energy blockages, raise vibrational frequency, and facilitate spiritual growth.
4. **Visualization and Intentional Focus**: Visualization techniques, where individuals focus their mind on specific images, sensations, or ideas, can also help shift vibrational frequency. By holding a clear mental image of a desired state—whether that is peace, love, or spiritual connection—individuals can attune their energy to the frequency of that state and thus access new dimensions of experience.

14.3 Mental Dimensions: The Power of Consciousness

How the Mind Acts as a Portal to Other Dimensions

One of the most fascinating aspects of **human consciousness** is its ability to act as a **portal** to other dimensions. While the physical senses are limited in their ability to perceive reality, the **mind** has the capacity to transcend these limitations, accessing dimensions of thought, imagination, and experience that exist beyond the material world.

In many spiritual and philosophical traditions, the mind is viewed as a bridge between the **physical world** and **higher realms** of existence. It is through the mind that we dream, imagine, and create realities that do not yet exist. This capacity for **imaginative thought** suggests that consciousness itself is not confined to the material dimension but can travel freely through mental dimensions, exploring new ideas, possibilities, and experiences.

States of Consciousness as Gateways to New Dimensions

One of the primary ways in which the mind accesses other dimensions is through **altered states of consciousness**. These states, which can be induced through practices such as **meditation, lucid dreaming**, or the use of **psychedelic substances**, allow individuals to temporarily step outside of ordinary reality and explore new dimensions of existence.

- **Meditation**: In deep meditation, individuals often report experiencing **expanded awareness**, where the boundaries of the physical self dissolve, and they feel connected to a larger, universal consciousness. This state is often described as a gateway to higher dimensions, where the limitations of time and space no longer apply.

- **Lucid Dreaming**: In a **lucid dream**, the dreamer becomes aware that they are dreaming and can consciously interact with the dream world. This state of heightened awareness allows individuals to explore **alternate realities** where the normal rules of physics and logic do not apply. Lucid dreamers often describe these experiences as journeys into **mental dimensions**, where thought shapes reality and the dreamer has control over the unfolding narrative.
- **Psychedelic Experiences**: The use of **psychedelic substances** has been practiced for thousands of years in cultures around the world as a way to access higher dimensions of consciousness. These substances, such as **ayahuasca, psilocybin**, or **LSD**, can induce powerful altered states where individuals report experiencing **ego dissolution**, encounters with non-physical entities, and the perception of new dimensions that exist beyond the material world.

The Role of Imagination and Thought in Shaping Personal Dimensions of Reality

At the heart of the mind's ability to access other dimensions is the power of **imagination**. Human beings have the unique capacity to **create** realities in their minds, whether through **fantasy, daydreaming**, or **creative problem-solving**. This imaginative ability suggests that consciousness is not constrained by the physical world but can explore **infinite dimensions** of possibility.

In many spiritual traditions, thought is seen as a **creative force** that shapes reality. The idea that **thought creates reality** is central to practices such as **affirmations, visualization**, and **intention setting**, where individuals use the power of their mind to manifest desired outcomes in the physical world. According to this view, the dimensions we experience are not fixed but are shaped by the **beliefs, thoughts**, and **intentions** we hold in our minds.

- **Example**: In the practice of **law of attraction**, individuals are taught to focus their thoughts and intentions on what they want to create in their lives, whether it is love, success, or spiritual growth. By holding a clear mental image of the desired outcome, they align their vibrational frequency with that reality, allowing it to manifest in their physical experience.

In this sense, the mind is a **dimensional tool**, capable of shaping the realities we experience by directing our attention and energy toward specific outcomes. The dimensions we inhabit are not static but are constantly being shaped by the thoughts and beliefs we hold.

14.4 Spiritual Dimensions: Connecting with the Higher Self and Universal Energy

The Spiritual Belief in Higher Planes of Existence

Many spiritual traditions throughout history have taught that there are **higher planes of existence** beyond the physical world. These **spiritual dimensions** are often described as **subtle realms** that exist in parallel to our material reality but operate according to different laws and principles.

- **The Astral Plane**: One of the most commonly referenced spiritual dimensions is the **astral plane**, a realm of existence that is said to lie between the physical world and

higher spiritual realms. In many esoteric traditions, the astral plane is considered a place where souls travel after death or during **out-of-body experiences**. It is also seen as a realm where beings such as **angels**, **spirit guides**, and **ascended masters** reside.
- **The Etheric Plane**: Another spiritual dimension is the **etheric plane**, which is believed to be the subtle energy field that surrounds and permeates the physical body. The etheric plane is often associated with **energy healing** and **chakra work**, as it is thought to be the dimension where energy blockages manifest before appearing in the physical body.

These spiritual dimensions are not perceived through ordinary senses but are accessed through **spiritual practices** such as **meditation**, **shamanic journeying**, and **dreamwork**. Those who explore these dimensions often describe them as realms of **light**, **love**, and **higher wisdom**, where the limitations of the physical world fall away and deeper truths about existence are revealed.

Exploring Ancient Traditions That Access These Dimensions

Many ancient traditions have developed practices specifically designed to access spiritual dimensions and connect with **higher realms of consciousness**. These practices are often deeply rooted in the **mythology**, **cosmology**, and **spiritual beliefs** of the cultures in which they originate.

- **Shamanic Journeys**: In indigenous shamanic traditions, the **shaman** acts as a bridge between the physical and spiritual worlds. Through the use of **drumming**, **chanting**, and **plant medicines**, shamans enter altered states of consciousness, allowing them to travel to spiritual dimensions and communicate with **spirits**, **ancestors**, and **deities**. These journeys are often undertaken for the purpose of **healing**, **guidance**, or **protection**.
- **Tibetan Meditation Practices**: In Tibetan Buddhism, practitioners engage in **meditation** and **visualization** techniques to access higher spiritual dimensions and connect with **bodhisattvas**, **deities**, and **enlightened beings**. The practice of **phowa**, for example, is a meditation technique designed to help the soul transfer to higher realms after death. Tibetan meditation practices also emphasize the importance of recognizing the **illusory nature of reality**, which can lead to direct experiences of higher dimensions.
- **Mystical Experiences and Visions**: Throughout history, mystics and spiritual seekers have reported experiences of **visions**, **revelations**, and **mystical encounters** that take them beyond the confines of the physical world. These experiences are often described as **glimpses** into higher dimensions, where individuals feel a profound sense of **unity**, **love**, and **connection** with the **divine**.

Vibrational Beings and Entities in Other Dimensions

One of the most intriguing aspects of spiritual dimensions is the possibility that these realms are inhabited by **vibrational beings** or **conscious entities**. These beings are said to exist at **higher vibrational frequencies** than our physical dimension, making them invisible to ordinary perception.

- **Angels and Spirit Guides**: In many spiritual traditions, angels and spirit guides are believed to reside in higher dimensions, offering guidance, protection, and support to

those in the physical world. These beings are often described as beings of **pure light** or **energy**, whose presence can be felt through subtle vibrations, signs, or synchronicities.
- **Ascended Masters**: In certain esoteric traditions, ascended masters are beings who have transcended the cycle of **birth** and **death** and now exist in higher dimensions. These masters are believed to offer wisdom and spiritual guidance to those seeking enlightenment, often appearing in dreams, meditations, or visions.
- **Extraterrestrial Beings**: Some spiritual and metaphysical traditions propose that extraterrestrial beings, or **interdimensional entities**, may also exist in higher dimensions. These beings are thought to possess advanced knowledge and technology, which allows them to travel between dimensions and communicate with humans.

14.5 The Akashic Records: A Dimension of Universal Knowledge

Exploring the Concept of the Akashic Records

The **Akashic Records** are a concept found in many **esoteric** and **spiritual traditions**, particularly within **theosophy** and **Hindu philosophy**. The term "Akasha" comes from the **Sanskrit** word meaning **ether** or **sky**, and the Akashic Records are believed to be a metaphysical dimension that contains the entire **history of the universe**, including every thought, action, and event that has ever occurred or will occur.

According to this belief, the Akashic Records are a **repository of universal knowledge**, where the past, present, and future are all recorded. Spiritual seekers, mystics, and psychics claim to access the Akashic Records through **meditation**, **trance states**, or **divination**, using this dimension to gain insights into personal and collective history, spiritual growth, and **soul purpose**.

Accessing the Akashic Dimension Through Spiritual Practice

Accessing the **Akashic dimension** requires specific spiritual practices designed to attune one's consciousness to this subtle realm of knowledge. These practices may include:

- **Guided Meditation**: Many practitioners use **guided meditations** that lead them into a deep state of relaxation, opening the mind to access the Akashic Records. During these meditations, individuals may visualize themselves entering a **sacred space** or **library** where they can review the records of their soul's journey.
- **Prayer or Invocation**: In some traditions, prayers or **invocations** are used to request permission to enter the Akashic Records. These prayers are intended to establish a connection with the **guardians** of the records, who are believed to oversee access to this dimension.
- **Energy Healing**: Some practitioners of **Reiki** or other forms of **energy healing** claim to access the Akashic Records during healing sessions, using this dimension to gain insights into the underlying spiritual causes of physical or emotional issues.

14.6 Techniques for Accessing Hidden Dimensions

Meditation and Breathwork for Altering States of Consciousness

The most common technique for accessing hidden dimensions is **meditation**, which allows individuals to quiet the mind, focus inward, and attune to higher levels of consciousness. Through sustained practice, meditators can raise their vibrational frequency, opening the door to experiences beyond the physical world.

- **Mindfulness Meditation**: Practitioners focus on their breath or present-moment awareness, training the mind to remain calm and centered. This creates the mental space necessary for higher-dimensional experiences to emerge.
- **Pranayama**: In yoga, pranayama refers to the regulation of breath. This ancient practice, used for thousands of years, alters one's energy field and vibrational state, making it easier to access altered states of consciousness and higher dimensions.

Sound and Vibration Therapy for Dimensional Travel

Sound has long been used as a tool for altering consciousness and accessing higher dimensions. **Sound therapy** employs **frequency** and **vibration** to attune the body and mind to higher states of being, often using tools like **Tibetan singing bowls, tuning forks**, or **binaural beats**.

Rituals, Visualization, and Intentional Focus

Intentional rituals and **visualization techniques** allow practitioners to focus their consciousness on specific dimensions or entities. By holding a clear intention and visualizing the desired outcome, individuals can align their energy with higher realms, allowing for **dimensional travel** or spiritual encounters.

Technology's Emerging Role in Helping Humans Access Different States of Mind and Vibration

Emerging technologies, such as **virtual reality** and **brainwave entrainment**, are being developed to help individuals enter altered states of consciousness more easily. These technologies hold the potential to **bridge the gap** between the physical and spiritual dimensions, allowing for more controlled and repeatable explorations of higher realms.

14.7 What Lies Beyond: Unexplored Dimensions Waiting for Human Perception

As humanity continues to push the boundaries of science, consciousness, and spiritual exploration, the potential for discovering **unexplored dimensions** is vast. The next frontier of human evolution may lie in our ability to perceive and interact with dimensions that have so far been hidden from us.

The Potential for New Discoveries in Vibrational Science, Consciousness, and Spiritual Realms

As scientific understanding of **vibration**, **quantum physics**, and **consciousness** evolves, it is likely that new dimensions of reality will be uncovered, offering new insights into the nature of existence.

Future Technologies That May Allow Us to Explore These Dimensions More Fully

The development of **quantum technologies**, **neuroscience tools**, and **virtual reality platforms** may soon enable humanity to explore and interact with higher dimensions in ways that were previously unimaginable.

Humanity's Journey Toward Expanding Perception and Tapping into the Full Multiverse of Existence

The exploration of extra dimensions represents a profound journey of **self-discovery**, as humanity continues to expand its perception and embrace the full **multiverse** of existence.

Chapter 15: The Perceptual Multiverse and the Human Experience

15.1 How Perception Shapes Reality in the Multiverse

The Power of Perception

At the core of the **Perceptual Multiverse** is the idea that **perception** fundamentally shapes our experience of reality. The **multiverse theory**, traditionally applied in physics to explain the existence of multiple, potentially infinite, universes, is here adapted to the realm of **human thought** and **belief systems**. In this expanded context, each individual's **perception of reality** is influenced by a unique combination of **personal experiences**, **cultural conditioning**, **psychological factors**, and **spiritual beliefs**, which effectively creates their own personal "universe."

Perception is the lens through which we view and interpret everything around us. From simple sensory data—such as the way we experience **color**, **sound**, or **temperature**—to complex emotional or ideological interpretations—such as how we understand **morality**, **politics**, or **social justice**—our perceptions shape not only how we interpret the world, but how we react to it. This idea leads to a profound conclusion: reality, as experienced by individuals, is **subjective** and **fluid**, varying widely depending on the **filters** through which it is processed.

This concept is vital to understanding the Perceptual Multiverse, where **multiple realities** exist simultaneously. Different people can inhabit completely different realities while occupying the same physical space, and their interactions are colored by the **perceptual dimensions** in which they exist.

Perception as Reality Creation

In the Perceptual Multiverse, the idea that "**perception creates reality**" is more than a metaphor—it is a literal description of how reality unfolds. The thoughts we hold, the stories we tell ourselves, and the beliefs we subscribe to form the boundaries of our personal reality. For instance, individuals who believe in a **scarcity mindset**—where resources are limited and hard to come by—will experience a reality in which they constantly face competition, lack, and struggle. On the other hand, those who adopt an **abundance mindset** may perceive opportunities and wealth flowing toward them, even in the same external circumstances.

The **mind** acts as a **reality filter**—selecting, organizing, and interpreting data in ways that reinforce existing beliefs and expectations. This explains why two people can witness the same event and come away with entirely different interpretations. One person may see a world filled with danger and chaos, while another sees opportunities for growth and transformation.

When scaled up, the concept of perception as reality creation explains the existence of **collective realities** or **societal dimensions**—shared frameworks of understanding that are adopted by groups of people. Examples include **religions**, **political ideologies**, and **cultural**

norms. These shared realities are formed through the combined perceptions of individuals, reinforced by **social narratives** and **institutions** that create a sense of common truth.

- **Example**: Consider two individuals living in the same country. One believes in a narrative of national decline, focusing on political corruption, economic downturns, and social discord. Another believes in a narrative of national progress, highlighting technological innovation, economic growth, and social movements for equality. Though they inhabit the same physical space, their experiences of the country are shaped by the particular **dimension** of reality they choose to perceive.

The Influence of Thought and Belief Systems

Belief systems—whether philosophical, spiritual, or ideological—are a major force in shaping the **dimensions** individuals experience within the Perceptual Multiverse. Once someone subscribes to a particular belief system, they filter all incoming information through that lens, reinforcing their perception of reality and further anchoring themselves in that dimension.

For example, consider someone who adopts a **conspiracy theory** that the government is constantly monitoring and manipulating people. This belief will color how they interpret everyday events, from seeing surveillance cameras on street corners to reading news headlines. In their perceptual reality, these seemingly ordinary elements take on new meaning, confirming their belief in a **shadowy dimension** of control and surveillance.

In contrast, someone who believes in the benevolent nature of their government may see the same surveillance cameras as safety measures designed to protect the public. These two individuals inhabit **parallel realities**, both valid within their own perceptual dimensions, but drastically different in how they interpret and engage with the world.

Cognitive Filters and the Distortion of Reality

Cognitive filters play a critical role in the formation of perception. These filters—shaped by **psychology**, **culture**, and **personal experience**—determine how we interpret incoming information. Common cognitive filters include:

- **Confirmation Bias**: The tendency to seek out and prioritize information that confirms pre-existing beliefs while ignoring contradictory evidence. This filter helps reinforce the boundaries of a dimension by making it difficult to question or challenge core assumptions.
- **Selective Attention**: Our brains are constantly bombarded with sensory input, but we only pay attention to a small fraction of it. **Selective attention** means that we filter out irrelevant details to focus on what matters most to us, based on our current goals and beliefs. This process allows us to construct reality based on what we deem important, even if others see different aspects of the same situation.
- **Framing**: How information is presented or "framed" influences how it is perceived. For instance, describing an economic policy as a "tax cut for the rich" will produce a very different emotional reaction than framing it as "incentivizing investment and job creation." Both descriptions may refer to the same policy, but the framing shifts the perceptual dimension in which it is understood.

These filters create what can be seen as **distortions** or **subjective realities**. In extreme cases, they lead to the phenomenon of **polarized realities**, where different groups of people not only disagree on interpretations but seem to live in completely different worlds. For example, in modern politics, one group may inhabit a dimension where climate change is a pressing existential threat, while another resides in a dimension where climate change is a hoax or at least vastly exaggerated. Both groups are so deeply embedded in their perceptual realities that finding common ground can feel impossible.

15.2 The Multiverse as a Framework for Understanding Conflict and Cooperation

The Root of Conflict in Parallel Dimensions

One of the most important applications of the Perceptual Multiverse is its ability to explain the **source of conflict** in human society. Much of the discord we observe—whether it is political, cultural, or personal—can be traced back to the fact that different individuals and groups inhabit **parallel perceptual dimensions**. Each dimension is shaped by its own **belief systems**, **values**, and **narratives**, which often come into conflict with the dimensions of others.

In these situations, conflicts arise not because one side is objectively "right" and the other is "wrong," but because each side is perceiving reality through a different set of filters. The **multiverse framework** helps explain why seemingly intractable conflicts persist—people are literally living in **different perceptual universes**. From their perspective, their beliefs and actions make sense given the logic of the dimension they inhabit.

For example, consider the **culture wars** surrounding issues like **abortion**, **immigration**, or **gun control**. These debates often seem impossible to resolve because the two sides are operating within entirely different dimensions of thought. One side may see the issue in terms of **individual rights** and **freedom**, while the other side views it as a matter of **public safety** or **moral responsibility**. Both are valid within their respective perceptual realities, but reconciling them requires crossing dimensional boundaries—something that can be psychologically and emotionally difficult.

The Potential for Cooperation Between Dimensions

Despite the challenges posed by the existence of parallel dimensions, the **multiverse framework** also offers a hopeful vision for **cooperation**. By recognizing that others inhabit different perceptual realities, individuals and groups can begin to approach conflict with greater **understanding** and **empathy**. Instead of dismissing opposing views as irrational or misguided, the multiverse framework encourages us to see them as part of a **different dimension**—valid within its own context but shaped by different experiences, beliefs, and priorities.

One of the keys to cooperation between dimensions is the willingness to engage in **dialogue** and explore the **common ground** that exists across different realities. While two individuals may hold opposing views on a specific issue, they may share underlying values—such as the desire for safety, prosperity, or justice. By focusing on these shared values, individuals from different dimensions can begin to **co-create** solutions that address both perspectives.

- **Example**: In the realm of environmental policy, individuals who prioritize **economic growth** and those who prioritize **climate protection** may seem to inhabit opposing dimensions. However, they could potentially find common ground by supporting policies that promote **green technology** and **sustainable energy**, which would benefit both the economy and the environment. The multiverse framework allows for the possibility of such **synergistic solutions**, as it encourages us to look for **overlapping realities** rather than focusing solely on points of conflict.

Building Bridges Across Dimensional Boundaries

Building bridges between perceptual dimensions requires **openness**, **flexibility**, and **curiosity**. Individuals who are willing to explore perspectives outside of their own dimension are more likely to foster cooperation and mutual understanding. However, this process often involves **crossing boundaries** that feel uncomfortable or even threatening.

One of the ways to build bridges across dimensions is through the practice of **empathy**—the ability to see the world from another person's perspective. Empathy allows us to momentarily enter the perceptual dimension of another, gaining insight into the values, experiences, and fears that shape their reality. By doing so, we can better understand the **logic** of their worldview, even if we do not share it.

Another important tool for building bridges is **education**. When individuals are exposed to **diverse perspectives**, whether through formal education, travel, or conversation, they are more likely to recognize the existence of **multiple dimensions**. Education can broaden our perceptual horizon, helping us to see beyond the limitations of our own dimension and appreciate the complexity of the multiverse.

Lastly, the concept of **shared goals** is critical for cooperation across dimensions. While people may inhabit different realities, they often share similar **aspirations** for themselves, their families, and their communities. By identifying these shared goals, individuals and groups can work together to achieve **mutually beneficial outcomes**.

15.3 Living in Parallel Realities: Embracing the Multiverse in Everyday Life

The Benefits of Recognizing the Multiverse in Daily Experience

Understanding the existence of parallel realities within the Perceptual Multiverse has profound implications for how we navigate everyday life. Recognizing that we each inhabit our own **dimension of perception** encourages a more **open-minded** and **compassionate** approach to interactions with others. Rather than assuming that everyone shares the same understanding of reality, we can appreciate that others may see the world in fundamentally different ways.

This recognition can help reduce **frustration** and **misunderstanding** in our personal relationships, as it allows us to approach disagreements with the knowledge that the other person may be coming from a completely different perceptual framework. Instead of trying to "win" arguments or force others to adopt our view, we can focus on finding ways to **bridge the gap** between dimensions.

In addition, embracing the Perceptual Multiverse encourages **self-reflection**. By recognizing that our reality is shaped by our perceptions, we become more aware of the filters and biases that influence our understanding of the world. This self-awareness can lead to personal growth, as we learn to question our assumptions and explore new dimensions of thought.

- **Example**: A person who recognizes that they have been operating from a **fear-based dimension** may decide to shift their focus toward a **love-based dimension**, choosing to interpret events and interactions through a lens of compassion and trust rather than suspicion and anxiety.

Navigating Parallel Realities in Social and Political Life

On a broader scale, understanding the Perceptual Multiverse helps us navigate the **complex social and political landscape** of the modern world. In an era of increasing polarization and division, it is easy to become entrenched in one's own dimension, dismissing those who hold opposing views as ignorant, immoral, or irrational. However, the multiverse framework invites us to see these divisions not as evidence of a broken society but as a natural consequence of living in a world where **multiple realities coexist**.

By adopting this perspective, we can engage more **productively** in social and political discourse. Instead of viewing opposing viewpoints as threats to our own reality, we can approach them as opportunities to learn about **alternate dimensions**. This doesn't mean we have to agree with every viewpoint we encounter, but it does mean we can approach discussions with **curiosity** rather than hostility.

- **Example**: A political debate about immigration might be framed as a clash between those who prioritize **national security** and those who emphasize **human rights**. Instead of assuming that one side is right and the other is wrong, the multiverse framework encourages us to explore how each dimension is shaped by different values, experiences, and concerns. This understanding can lead to more nuanced discussions and the possibility of finding **compromises** that address the needs of both dimensions.

Developing Flexibility in Dimensional Thinking

One of the most valuable skills in the Perceptual Multiverse is the ability to **move fluidly** between different dimensions of thought. This **flexibility** allows individuals to navigate the complexities of modern life with greater ease, as they can adapt their perceptions to fit the situation at hand.

For example, someone who is adept at dimensional thinking may be able to switch between a **spiritual dimension** when engaging in personal reflection or meditation, a **scientific dimension** when analyzing data or solving problems, and a **social dimension** when interacting with friends or colleagues. This fluidity enables them to **integrate multiple perspectives**, leading to a richer and more balanced experience of reality.

Developing dimensional flexibility requires **practice** and **self-awareness**. It involves recognizing when we are locked into a particular perceptual dimension and making a conscious effort to **broaden our perspective**. This could involve stepping outside of our comfort zone, exposing ourselves to new ideas, or simply being open to the possibility that our current understanding of reality is not the only valid one.

- **Example**: A person who identifies strongly with their political beliefs may choose to explore media or literature from an opposing perspective as a way of broadening their dimensional understanding. This doesn't mean they have to abandon their beliefs, but it allows them to see how different realities are constructed and experienced.

The Multiverse as a Tool for Personal Growth

Finally, embracing the Perceptual Multiverse can be a powerful tool for **personal growth**. By recognizing that our reality is shaped by our perceptions, we gain the ability to **consciously create** the reality we want to experience. This involves taking ownership of our thoughts, beliefs, and actions, and aligning them with the **dimension** we wish to inhabit.

For example, someone who desires more **peace** and **joy** in their life might begin by examining the thoughts and beliefs that are keeping them anchored in a dimension of **stress** or **negativity**. By shifting their focus toward gratitude, mindfulness, and self-compassion, they can raise their **vibrational frequency** and move into a dimension where peace and joy are more readily accessible.

This process of dimensional shifting is not about escaping reality but about recognizing the power we have to shape our experience of the world. In this sense, the Perceptual Multiverse offers a path toward **greater empowerment**, as individuals learn to navigate the many dimensions of life with intention, flexibility, and awareness.

Conclusion: Embracing the Perceptual Multiverse

The concept of the Perceptual Multiverse offers a profound framework for understanding the human experience. By recognizing that we each inhabit our own dimension of reality—shaped by our perceptions, beliefs, and cognitive filters—we can approach life with greater empathy, flexibility, and self-awareness.

The multiverse framework not only helps explain the conflicts and cooperation we observe in society but also provides a roadmap for personal growth and meaningful interaction in a world of parallel realities. Whether navigating political debates, personal relationships, or spiritual journeys, the ability to move fluidly between dimensions offers a richer, more integrated experience of life.

By embracing the Perceptual Multiverse, we open ourselves to the vastness of human experience, recognizing that reality is not fixed but is instead a dynamic and multi-faceted landscape shaped by our thoughts, beliefs, and interactions. In doing so, we gain the power to create and inhabit the dimensions that best serve our highest potential.

Chapter 16: Disasters and Dimensional Synchronization

16.1 The Nature of Dimensional Convergence During Disasters

How External Events Act as Unifying Forces

In the **Perceptual Multiverse**, individuals often occupy different **dimensions of perception**, shaped by their unique experiences, beliefs, and cultural backgrounds. These dimensions, which can be political, social, or personal, create fragmented realities in which people interpret the world differently. However, certain **external events**—particularly disasters or crises—have the power to act as **unifying forces**, temporarily aligning these parallel realities into a **shared dimension**.

Disasters, whether natural (such as **hurricanes**, **earthquakes**, or **pandemics**) or human-made (such as **wars** or **economic collapses**), often serve as moments of **dimensional convergence**. In times of crisis, the differences that normally divide people—ideological, religious, or political—take a backseat to the immediate reality of survival, safety, and recovery. These events create a situation where individuals who previously inhabited vastly different perceptual dimensions are brought into a **singular shared reality** that demands collective attention.

- **Example**: After the September 11, 2001, terrorist attacks in the United States, people from diverse political, social, and religious backgrounds experienced a brief period of **dimensional synchronization**. Despite their differences, Americans, and even global observers, were temporarily unified by the shared reality of the attack's magnitude and the need for collective healing. In that moment, the **multiverse** of fragmented realities collapsed into one where the primary focus was on grief, solidarity, and response.

Disasters, therefore, have the ability to create a **dimensional field** where **individual interests** and **personal differences** are subordinated to the **greater need** for cooperation, survival, and collective action. During these moments, the boundaries between personal realities become porous, and people enter into a **shared perceptual dimension** that is shaped by the crisis at hand.

Understanding Reality Synchronization Through Crisis

The process of **reality synchronization** during a disaster can be understood as the **alignment of focus** across different dimensions. In everyday life, individuals operate in divergent realities, prioritizing their personal goals, beliefs, and narratives. However, during a crisis, the immediate threat or external challenge forces people to shift their focus toward a common goal—whether that is survival, rescue, or recovery.

This **collective focus** acts as a **dimensional attractor**, pulling individuals out of their personal perceptual dimensions and into a **synchronized reality**. In this shared dimension, the external crisis becomes the central organizing force, and individuals adjust their behaviors, thoughts, and priorities to align with the needs of the collective. In many ways, the crisis acts as a **gravitational force**, collapsing the multiplicity of realities into a single shared experience that transcends individual differences.

- **Example**: During natural disasters like hurricanes or wildfires, communities often come together in ways that defy normal social boundaries. Neighbors who may have had little in common before the disaster suddenly find themselves working side by side to help each other evacuate, find shelter, or rebuild. The **shared threat** of the disaster synchronizes their realities, creating a temporary dimension where **survival** and **mutual aid** become the dominant themes.

In this way, disasters serve as **dimensional gateways**, temporarily breaking down the barriers that separate individual perceptual dimensions and creating a shared space where cooperation and collective focus take precedence.

16.2 Disasters as Dimensional Gateways

The Mechanism of Reality Alignment During Collective Crises

The mechanism through which **reality alignment** occurs during disasters can be explained through the concept of **shared focus**. In normal circumstances, individuals are free to direct their attention toward a multitude of personal concerns—such as work, relationships, or ideological debates. However, during a disaster, the urgency of the situation commands the attention of everyone involved, creating a powerful **dimensional pull** that aligns otherwise divergent realities.

In this process, the **external event**—whether it is a **natural disaster**, **pandemic**, or **conflict**—serves as the anchor for **dimensional convergence**. The shared experience of the event overrides the individual's usual filters and perceptions, forcing them to engage with a reality that is dictated by the collective experience. As a result, individuals temporarily set aside their personal beliefs, priorities, and ideological frameworks in favor of focusing on the **common threat** or **shared goal**.

- **Example**: In the aftermath of a major earthquake, individuals from different socioeconomic, political, and religious backgrounds often find themselves participating in the same efforts to rescue survivors, provide medical aid, or rebuild damaged infrastructure. In these moments, the disaster creates a **unifying reality** where traditional social distinctions become irrelevant. The focus on immediate survival and recovery collapses the **multiverse of personal concerns** into a **singular dimension** of action and response.

This phenomenon can be thought of as a **dimensional collapse**—where the normally separate dimensions of thought, belief, and perception converge into a unified reality. The disaster creates a moment of **dimensional clarity**, where the complexity and fragmentation of normal life give way to a simplified, collective reality centered on the shared experience of the crisis.

Shared Interests and Priorities in the Face of Danger

One of the key factors that facilitate **dimensional convergence** during disasters is the emergence of **shared interests and priorities**. In ordinary life, individuals may prioritize different values, such as **individual freedom**, **economic security**, or **environmental sustainability**. However, during a crisis, these differences are overshadowed by the universal need for **safety**, **security**, and **survival**.

The alignment of shared interests creates a **temporary dimension** where individuals, regardless of their background or personal beliefs, come together to achieve common goals. These goals—such as finding food, water, shelter, or medical care—become the dominant organizing principles of the shared reality. In this unified dimension, the usual divisions that separate people are replaced by a collective sense of urgency and mutual dependence.

- **Example**: During the COVID-19 pandemic, people across the globe were forced to confront the same **public health crisis**, regardless of their individual circumstances or beliefs. The shared threat of the virus led to widespread cooperation in the form of **social distancing**, **mask-wearing**, and **vaccination efforts**. Although there were significant variations in how different communities responded to the crisis, the pandemic created a **shared dimension** where public health became a central concern for all.

The power of disasters to unify people lies in their ability to **transcend individual differences** and create a reality where collective survival becomes the overriding priority. In this way, disasters act as **dimensional gateways**, allowing people to step outside of their personal realities and enter a **shared dimension** where cooperation and mutual aid are essential.

How the Multiverse Temporarily Collapses into a Unified Dimension

In the Perceptual Multiverse, individuals typically inhabit parallel realities that are shaped by their unique perspectives, beliefs, and experiences. However, during disasters, these parallel realities can **collapse into a unified dimension**, where the distinctions between personal and collective experience blur. This **temporary collapse** occurs because the disaster creates a **common external reality** that demands collective attention and action.

The concept of **dimensional collapse** can be likened to a temporary **merging of timelines** or realities, where individuals who normally operate in separate dimensions are drawn together by the same external force. In these moments, the boundaries between different perceptual dimensions become less rigid, allowing for a greater degree of **interconnection** and **shared experience**.

- **Example**: After a major flood, people from different parts of a city may come together to help with evacuation efforts, distribute food and water, and assist in the rebuilding process. In this shared dimension, the usual divisions between neighborhoods, social classes, or political affiliations become irrelevant. The flood creates a reality where everyone is focused on the same goal: surviving and recovering from the disaster.

This **dimensional collapse** is temporary, lasting only as long as the external crisis remains the primary focus of attention. Once the disaster has passed and the immediate threat has subsided, individuals begin to **reassert their personal realities**, leading to a gradual return to the fragmented multiverse of parallel dimensions.

16.3 Traveling Between Realities: Collective Experiences as Gateways

How Collective Experiences Facilitate Dimensional Shifts

Disasters and collective crises provide a unique opportunity for **dimensional travel**—the process of moving between different perceptual realities. When individuals experience a crisis together, they are temporarily drawn into a shared reality, but this experience can also serve as a **gateway** for more permanent dimensional shifts.

Collective experiences, particularly those that involve intense **emotional** or **physical challenges**, can act as catalysts for **dimensional expansion**. In the aftermath of a disaster, individuals may find that their perspectives on life, community, and society have fundamentally changed. The crisis may have forced them to confront new realities, and in doing so, they may find themselves inhabiting a different **dimension of thought** than they did before the event.

- **Example**: After surviving a natural disaster, an individual who previously lived a life focused on material success and personal achievement may experience a profound shift in priorities. The shared experience of the disaster may lead them to place greater value on **community**, **relationships**, and **spiritual growth**, as they come to realize the fragility of life and the importance of collective well-being. In this sense, the disaster acts as a **gateway** to a new dimension of perception, one that is more aligned with **interconnectedness** and **mutual support**.

Collective experiences, especially those marked by hardship or struggle, have the power to **collapse** the boundaries between individual realities, creating a space where new perspectives and values can take root. These experiences facilitate **dimensional shifts** by exposing individuals to new ways of thinking and living, often leading to lasting changes in their perception of reality.

The Role of Shared Focus in Collapsing Multiple Realities into One

A key factor in the **collapse of multiple realities** during a disaster is the role of **shared focus**. When a large group of people is collectively focused on the same event or goal, the usual fragmentation of the multiverse gives way to a **unified dimension** where the distinctions between individual realities blur. This collective focus acts as a kind of **dimensional glue**, holding people together in a shared reality that is shaped by the crisis.

In times of crisis, the shared focus on **survival**, **safety**, or **recovery** overrides the normal concerns of daily life. Individuals who might normally occupy different dimensions of thought—based on their political views, religious beliefs, or personal interests—are brought together by the need to address the immediate challenges posed by the disaster. In this unified dimension, the external reality of the crisis becomes the dominant organizing force, and individuals align their actions and priorities accordingly.

- **Example**: During the early days of the COVID-19 pandemic, the shared focus on **public health** and **safety** led to a temporary collapse of multiple realities into a singular dimension where issues such as **mask mandates**, **social distancing**, and **vaccination** became central to daily life. People who might have had different priorities before the pandemic were suddenly united by the shared experience of navigating the health crisis.

The power of shared focus lies in its ability to create **temporary dimensional coherence**, where individuals from different walks of life come together to address a common challenge. This coherence allows for the **alignment** of thoughts, actions, and priorities, creating a unified dimension that is shaped by the collective experience of the crisis.

16.4 Gradual Return to Individual Realities

The Post-Disaster Fragmentation of Shared Dimensions

While disasters have the power to temporarily collapse parallel realities into a unified dimension, this convergence is often short-lived. Once the immediate crisis has passed, individuals begin to **return to their personal realities**, and the **multiverse** of fragmented dimensions reasserts itself. This **post-disaster fragmentation** marks the end of the shared dimension created by the crisis and the return to a more **divergent** set of realities.

The process of returning to individual realities is gradual. In the immediate aftermath of a disaster, people may still feel a sense of **solidarity** and **shared purpose**, but as the crisis fades into the background, the differences that separate individuals—such as political beliefs, cultural values, or personal priorities—begin to re-emerge. Over time, the **dimensional coherence** created by the disaster gives way to the **fragmentation** of the multiverse, as people reassert their personal perspectives and return to their normal routines.

- **Example**: After a natural disaster like a hurricane or earthquake, there is often a period of **community solidarity**, where neighbors help each other rebuild and recover. However, as the immediate need for cooperation diminishes, individuals gradually return to their own lives, and the sense of **collective unity** fades. People once again become focused on their personal concerns, and the shared dimension of recovery gives way to the re-establishment of parallel realities.

The post-disaster fragmentation of shared dimensions is a natural process, reflecting the tendency of individuals to prioritize their **personal realities** once the external crisis is no longer the primary focus. While the temporary convergence created by the disaster may have lasting effects, the return to normalcy often signals the reassertion of the **multiverse** of individual experiences and perceptions.

How Personal Beliefs and Perceptions Reassert Themselves Over Time

As individuals move away from the shared dimension created by the disaster, their **personal beliefs** and **perceptions** begin to reassert themselves. The external event that once unified their focus is no longer the dominant force shaping their reality, and they gradually return to the perceptual dimensions that are most aligned with their **core beliefs** and **values**.

This reassertion of personal realities can sometimes lead to **tension** or **conflict**, especially if the shared experience of the disaster has created new dimensions of thought that challenge previously held beliefs. For example, individuals who experienced a **spiritual awakening** or a shift in values during the disaster may find themselves at odds with others who have returned to their pre-crisis perspectives.

- **Example**: After surviving a natural disaster, an individual may develop a deeper sense of **community** and **interconnectedness**, placing greater value on helping others and fostering social bonds. However, as life returns to normal, they may encounter resistance from friends or family members who have reverted to their previous focus on **individualism** or **material success**. This can create a sense of **disconnection**, as the individual struggles to reconcile their new dimensional perspective with the realities of those around them.

The process of **reasserting personal beliefs** and returning to individual realities is not always smooth, and it often involves a period of **adjustment** as individuals navigate the tension between their shared experience of the disaster and the return to their own perceptual dimensions.

16.5 Empathy and Shared Experience as Bridges Between Dimensions

The Role of Empathy in Creating Temporary Dimensional Convergence

One of the most powerful forces that facilitate **dimensional convergence** during disasters is **empathy**. When individuals experience a crisis together, they often develop a deep sense of empathy for one another, as they recognize the shared humanity that transcends their differences. This empathy acts as a bridge between **parallel realities**, allowing people to temporarily step outside of their own perceptual dimensions and enter into the experience of others.

Empathy fosters a sense of **connection** and **understanding**, creating a shared dimension where the focus is on mutual support and collective well-being. In times of disaster, empathy can break down the barriers that normally separate individuals, allowing for greater **cooperation** and **solidarity**.

- **Example**: During the aftermath of a natural disaster, people often go out of their way to help those who are suffering, regardless of their background or beliefs. The shared experience of the crisis creates a sense of **oneness** that transcends the usual social divisions, as individuals empathize with the pain and struggles of others.

Empathy not only facilitates **temporary dimensional convergence** but also has the potential to create lasting bridges between dimensions. By fostering a deeper understanding of others, empathy can help individuals remain connected to the shared dimension even after the crisis has passed, allowing for continued cooperation and mutual support.

Why Shared Experience Creates a Sense of Oneness Beyond Individual Beliefs

Shared experiences, especially those that involve **collective suffering** or **struggle**, have a unique ability to create a sense of **oneness** that transcends individual beliefs. In moments of crisis, people are often reminded of their common humanity and the fragility of life. This shared recognition can dissolve the boundaries between personal realities, creating a dimension where the focus is on **connection**, **compassion**, and **mutual aid**.

The sense of **oneness** that emerges during disasters is not based on shared beliefs or ideologies, but on the recognition of a shared experience. In these moments, the differences that normally separate people—such as political views, cultural values, or personal ambitions—become less important, as the focus shifts to the **common goal** of surviving and recovering from the crisis.

- **Example**: After a major earthquake, people from different religious or political backgrounds may come together to provide aid to survivors. In this shared dimension, the focus is not on their individual beliefs, but on the shared experience of responding to the disaster and helping those in need. This sense of **oneness** creates a temporary dimension where the usual divisions between people are replaced by a collective focus on healing and recovery.

16.6 Disasters as Catalysts for Dimensional Awareness

How Experiencing New Realities During Crises Expands Consciousness

Disasters, while often tragic, can also serve as **catalysts for dimensional awareness**. By forcing individuals to confront new realities and step outside of their usual perceptual dimensions, crises can lead to profound shifts in **consciousness**. The shared experience of the disaster may reveal aspects of reality that were previously hidden or ignored, expanding the individual's awareness of the **multiverse** and their place within it.

In many cases, the experience of surviving a disaster can lead to a deepening of **spiritual awareness** or a greater appreciation for the **interconnectedness** of all life. The crisis may challenge previously held beliefs, forcing individuals to reevaluate their priorities and consider new dimensions of thought and perception.

- **Example**: After surviving a near-death experience during a natural disaster, an individual may experience a shift in consciousness, leading them to explore **spiritual dimensions** or pursue a path of **personal growth**. The disaster acts as a **gateway** to new realities, expanding their awareness of the multiverse and their role within it.

By temporarily collapsing multiple realities into a shared dimension, disasters provide an opportunity for individuals to experience new dimensions of thought, belief, and perception. These experiences can lead to lasting changes in consciousness, as individuals integrate the lessons learned during the crisis into their everyday lives.

The Long-Term Impact of Temporarily Occupying a Shared Dimension

The long-term impact of temporarily occupying a shared dimension during a disaster can be profound. For many individuals, the experience of **dimensional convergence** creates a lasting sense of **connection** to others and a deeper awareness of the **multiverse**. This expanded consciousness may lead to changes in how individuals interact with the world, as they become more aware of the **interconnectedness** of all life and the importance of **collective action**.

In some cases, the shared experience of the disaster may inspire individuals to become more **involved in their communities**, to pursue **spiritual growth**, or to advocate for **social change**. The temporary collapse of parallel realities into a unified dimension can serve as a powerful reminder of the **potential for cooperation** and **mutual support**, even in the face of overwhelming challenges.

- **Example**: After surviving a major flood, an individual may be inspired to become more involved in **environmental activism** or **disaster relief efforts**, recognizing the importance of **community resilience** and **sustainability**. The shared experience of the disaster has expanded their awareness of the multiverse, leading them to pursue new dimensions of thought and action.

In this way, disasters can act as **catalysts for transformation**, expanding individual and collective consciousness and fostering greater **dimensional awareness**. By temporarily collapsing multiple realities into a shared dimension, crises create the conditions for profound shifts in perception, ultimately leading to personal and societal growth.

Conclusion: Disasters and Dimensional Synchronization

In the context of the **Perceptual Multiverse**, disasters serve as powerful **dimensional gateways**, temporarily collapsing parallel realities into a shared dimension where collective focus, empathy, and cooperation take precedence. While these moments of dimensional convergence are often short-lived, they have the potential to create lasting shifts in **consciousness** and expand our awareness of the multiverse.

By understanding how disasters synchronize realities and create shared experiences, we gain insight into the mechanisms of **dimensional travel**, the power of **collective focus**, and the potential for **personal growth** and **societal transformation**. Whether through the temporary collapse of parallel realities or the long-term impact of expanded dimensional awareness, disasters offer a unique opportunity to explore the **multiverse** and deepen our understanding of the interconnectedness of all life.

Conclusion: Understanding the Multiverse of Thought

The Implications of the Perceptual Multiverse Theory

The concept of the **Perceptual Multiverse** proposes that reality, as we experience it, is not a fixed or singular entity. Instead, it is fluid, multidimensional, and shaped by individual and collective perception. Each person's reality is deeply influenced by their **belief systems**, **thought patterns**, **cultural frameworks**, and **personal experiences**. As a result, the world we live in is not one monolithic reality, but rather a **multiverse** of **parallel dimensions**, each defined by the unique perspectives of the individuals and groups who inhabit them.

This theory extends beyond the purely **philosophical** or **metaphysical**, reaching into the realms of **science**, **psychology**, **sociology**, and even **politics**. By understanding that human beings live in **multiple realities** simultaneously, shaped by the dimensions they perceive and interact with, we can gain deeper insight into the complexities of **human behavior**, **conflict**, and **cooperation**. Recognizing this allows us to better navigate the challenges of **modern life**, from interpersonal relationships to societal debates, and opens the door to more **compassionate**, **flexible**, and **informed** approaches to life.

Reality as a Subjective Experience

One of the most profound implications of the Perceptual Multiverse Theory is the idea that **reality** is, to a large extent, a **subjective experience**. While we may all share a common physical world, our **interpretations** of that world vary dramatically based on the perceptual lenses through which we view it. These lenses are shaped by a range of factors, including:

1. **Cultural Influences**: The values, beliefs, and norms of a society play a significant role in shaping how individuals perceive reality. A person raised in a **collectivist culture** may prioritize community and cooperation, while someone from an **individualist culture** might emphasize personal freedom and self-reliance. These cultural differences create distinct **dimensional realities** that often clash in global discourse.
2. **Personal Experiences**: Every person's unique life experiences—such as **trauma, success, education**, and **relationships**—create a **personal dimension** in which their worldview is grounded. For example, someone who has experienced poverty may live in a dimension where **scarcity** is a primary focus, while a person who has always had financial security may inhabit a dimension where **abundance** is the norm.
3. **Belief Systems**: Religious, spiritual, and ideological beliefs shape how individuals interpret the world around them. A person who believes in a **spiritual multiverse** may perceive dimensions of reality that transcend the physical, while a materialist might restrict their perception to the tangible, measurable aspects of existence. These belief systems often lead to **parallel realities** that are difficult to reconcile without mutual understanding.
4. **Cognitive Frameworks**: Psychological theories, such as **cognitive bias**, help explain how the human mind processes information in ways that confirm pre-existing beliefs. **Confirmation bias**, **selective perception**, and **framing** are all tools that the mind uses to filter reality, creating a personalized dimension that reinforces one's worldview. These filters can either enhance or limit one's ability to perceive alternate dimensions of reality.

Recognizing that reality is **subjective** helps us understand why people often experience the same event or situation in radically different ways. Two individuals might witness the same political protest, for example, but one might see it as a justified demand for justice, while the other views it as a threat to societal order. Both interpretations are valid within their respective **dimensional realities**, but they are fundamentally different because they are filtered through distinct lenses of perception.

Dimensional Fluidity: Moving Between Realities

While most people tend to operate within a particular dimension of thought, the **Perceptual Multiverse Theory** also emphasizes the **fluidity** of these dimensions. People are not locked into a single reality; instead, they have the capacity to **shift** between dimensions based on new information, experiences, or changes in their mental or emotional state. This concept is crucial for understanding how personal growth, societal change, and even **conflict resolution** occur.

1. **Growth Through Experience**: As individuals encounter new experiences—whether through education, travel, or personal relationships—they often find themselves moving between **dimensional realities**. A person who has always lived in a homogeneous cultural environment might experience a dimensional shift when they travel to a country with different values, customs, and beliefs. This exposure expands their perception of reality, creating a more **complex and multidimensional** understanding of the world.

2. **Dimensional Shifts in Times of Crisis**: As explored in **Chapter 16**, disasters and crises can trigger **dimensional synchronization**, where individuals who normally inhabit different realities come together in a **unified dimension**. However, this convergence is often temporary, and once the crisis has passed, individuals tend to return to their **personal dimensions**. The fluid nature of reality is evident in these moments, as people move in and out of **shared dimensions** depending on the external circumstances.
3. **Resolving Conflict**: In situations of interpersonal or societal conflict, the ability to **shift between dimensions** is crucial for finding common ground. When two parties are entrenched in opposing realities, resolution becomes difficult unless one or both are willing to expand their perspective and explore the other's dimension of thought. For example, in political debates, a person may move between a **libertarian** and a **socialist** dimension, seeking to understand the values and concerns that define each reality. This fluidity allows for greater **empathy**, **compromise**, and **collaborative problem-solving**.
4. **Spiritual and Mystical Experiences**: The concept of **dimensional fluidity** is also central to spiritual practices and mystical experiences. Through **meditation**, **prayer**, or **altered states of consciousness**, individuals can access **higher dimensions** of thought that transcend their ordinary reality. These experiences often lead to profound shifts in perception, allowing individuals to see beyond the limitations of the physical world and into realms of **spiritual insight** or **universal connectedness**.

The Impact of Collective Realities on Society

Beyond the individual, the **Perceptual Multiverse Theory** has important implications for understanding the **collective realities** that shape entire societies. When groups of people share similar beliefs, values, and cultural narratives, they collectively create a **shared dimension**—a societal reality that influences how they interact with the world and with each other. These collective realities are powerful forces, as they define the **norms**, **laws**, and **institutions** that govern social life.

1. **Cultural Realities**: Cultures are built upon shared narratives and traditions that define what is considered normal, acceptable, and valuable. These cultural realities are a form of **collective dimension**, where individuals are socialized into specific ways of thinking and behaving. For example, in a culture that values **hierarchical authority**, individuals may experience a dimension where respect for elders and deference to leaders are central tenets. In contrast, a culture that values **egalitarianism** might foster a dimension where personal freedom and equality are paramount.
2. **Political Realities**: Political ideologies also create **collective dimensions** that shape the behavior and expectations of citizens. A society that operates within a **democratic dimension** may emphasize individual rights, participation in governance, and freedom of expression. Meanwhile, a society that operates within an **authoritarian dimension** may prioritize state control, order, and security. These political dimensions often come into conflict on the global stage, as each dimension seeks to impose its reality on others.
3. **Religious Realities**: Religions offer another form of **collective dimension**, where shared beliefs about the nature of the universe, morality, and the afterlife create a cohesive worldview for adherents. Religious dimensions often transcend physical reality, as they posit the existence of **spiritual planes**, **divine beings**, or **unseen forces** that shape the material world. For believers, these dimensions are just as real—if not more so—than the physical reality they inhabit.
4. **Economic Realities**: Economic systems also create distinct **dimensions of thought**, where the values of **capitalism**, **socialism**, or **mixed economies** define the boundaries

of what is possible or desirable. In a **capitalist dimension**, the pursuit of profit, individual entrepreneurship, and market competition are central to the experience of reality. In contrast, a **socialist dimension** may focus on collective ownership, wealth redistribution, and social welfare. These economic dimensions shape how individuals perceive opportunity, success, and fairness in society.

The **Perceptual Multiverse Theory** highlights the importance of recognizing and respecting these **collective dimensions**, as they offer insight into the **diverse realities** that exist within a globalized world. By understanding how these shared realities influence behavior, we can better navigate cross-cultural interactions, political debates, and economic negotiations.

Flashpoint Issues as Gateways to New Realities

What Are Flashpoint Issues?

Throughout the book, we have explored how **flashpoint issues**—highly contentious or emotionally charged societal debates—serve as **dimensional gateways** within the Perceptual Multiverse. Flashpoint issues often bring to the surface **deeply entrenched beliefs** and **values**, creating moments of intense conflict where multiple dimensions collide. However, these flashpoints also represent **opportunities** for growth, transformation, and the expansion of reality.

Flashpoint issues can be understood as **points of convergence** where individual and collective realities intersect. These issues are often divisive precisely because they touch on **core beliefs** and challenge individuals to either defend their existing dimension or explore new dimensions of thought.

- **Example**: The debate over **climate change** is a classic flashpoint issue that involves multiple dimensions of perception. On one side, there are those who view climate change as an existential threat that demands immediate action, often operating within a dimension where **environmental sustainability** and **global cooperation** are paramount. On the other side, there are those who see the issue as overblown or even a hoax, prioritizing **economic growth** and **personal freedom** over environmental concerns. The clash of these dimensions creates a flashpoint, but it also opens the door for individuals to expand their awareness and explore new realities.

Flashpoint issues can arise in a variety of contexts, including **political**, **social**, **environmental**, and **moral** debates. They are characterized by their ability to evoke strong emotional responses, as they challenge individuals to confront the boundaries of their perceptual dimensions and consider new ways of thinking.

The Transformative Potential of Flashpoint Issues

Flashpoint issues serve as powerful gateways to new realities because they force individuals to **question** their assumptions and **re-examine** their beliefs. While these issues often lead to conflict, they also create opportunities for **personal growth** and **dimensional expansion**. When individuals are exposed to **alternative viewpoints** or **new information** during a flashpoint, they may begin to shift their perception, leading to a **dimensional shift** that transforms their understanding of reality.

1. **Cognitive Dissonance and Growth**: Flashpoint issues often trigger **cognitive dissonance**, a state of psychological discomfort that arises when individuals are confronted with information that contradicts their existing beliefs. While cognitive dissonance can lead to **defensiveness** and **resistance**, it can also serve as a catalyst for **growth**. When individuals are willing to sit with the discomfort of dissonance and explore new ideas, they can experience a **dimensional breakthrough**, expanding their perception of reality.
2. **Emotional Intensity and Transformation**: The emotional intensity of flashpoint issues is part of what makes them such powerful gateways. When individuals are deeply invested in an issue—whether it's **reproductive rights**, **immigration policy**, or **racial justice**—the stakes feel incredibly high. This emotional investment can serve as a driving force for **transformation**, pushing individuals to explore new dimensions of thought and action. For example, someone who has always opposed immigration reform may experience a shift in perspective after hearing a personal story from an immigrant, allowing them to step into a new **dimension of empathy** and **compassion**.
3. **Collective Flashpoints and Societal Change**: Flashpoint issues are not only transformative on an individual level; they also have the potential to drive **societal change**. When a critical mass of people begins to shift their perception around a flashpoint issue, entire **collective dimensions** can change. This is often how social movements are born—when enough individuals experience a **dimensional shift**, their collective action can reshape the broader reality. Examples include the **civil rights movement**, the **LGBTQ+ rights movement**, and the **environmental movement**, all of which were sparked by flashpoint issues that transformed individual and collective realities.

Navigating Flashpoint Issues in the Multiverse

One of the greatest challenges of living in a **multiverse of thought** is navigating the flashpoint issues that arise within and between dimensions. Because flashpoint issues often evoke strong emotions and deeply held beliefs, they can lead to **polarization** and **entrenchment**, where individuals become more rigid in their views rather than open to new perspectives. However, by approaching flashpoint issues with **dimensional awareness**, we can begin to see them not as intractable conflicts, but as **opportunities** for personal and societal growth.

1. **Openness to Dimensional Shifts**: The first step in navigating flashpoint issues is to cultivate an openness to **dimensional shifts**. This means recognizing that our current dimension of thought is not the only valid reality, and being willing to explore new perspectives, even if they challenge our deeply held beliefs. By approaching flashpoint issues with curiosity rather than defensiveness, we create the conditions for **dimensional expansion**.
2. **Empathy as a Dimensional Bridge**: As discussed in **Chapter 16**, empathy plays a crucial role in bridging the gap between parallel realities. When we engage with others who hold opposing views on a flashpoint issue, empathy allows us to step into their dimension of thought and see the world from their perspective. This doesn't mean we have to agree with them, but it does mean we can approach the conversation with greater understanding and respect.
3. **Critical Thinking and Media Literacy**: In a world where **misinformation** and **disinformation** are rampant, critical thinking and media literacy are essential tools for navigating flashpoint issues. By questioning the sources of information we consume and engaging in thoughtful analysis, we can avoid falling into **dimensional echo chambers** and instead explore a broader range of perspectives.

4. **Collective Action for Dimensional Change**: Finally, flashpoint issues often require **collective action** to bring about dimensional change. Whether it's through activism, education, or policy reform, individuals who have experienced a dimensional shift around a flashpoint issue can work together to reshape the **collective reality**. By organizing around shared values and goals, people can co-create new dimensions of thought that reflect a more **equitable**, **just**, and **sustainable** future.

The Future of Humanity in a Multiverse of Infinite Dimensions

Embracing Dimensional Awareness for Personal and Collective Growth

As humanity continues to evolve, the concept of the **Perceptual Multiverse** offers a powerful framework for understanding the complexities of **individual experience**, **societal dynamics**, and **global challenges**. By recognizing that reality is not a singular, fixed entity, but rather a multiverse of **parallel dimensions**, we can begin to embrace **dimensional awareness** as a tool for personal and collective growth.

1. **Personal Empowerment**: On an individual level, embracing dimensional awareness allows us to take greater control over our own reality. By recognizing the power of **perception**, **belief**, and **thought**, we can shape the dimensions we inhabit and create the life we want to experience. This involves becoming more aware of the **filters** that shape our perception, questioning our assumptions, and cultivating a mindset of **growth** and **exploration**.
2. **Building Compassionate Societies**: On a societal level, dimensional awareness offers a path toward building more **compassionate** and **inclusive** societies. By recognizing that different groups of people inhabit different dimensions of thought, we can approach **cultural** and **political** conflicts with greater empathy and understanding. Instead of dismissing opposing views as irrational or wrong, we can see them as valid within their own dimensional framework, creating the possibility for **dialogue** and **collaboration**.
3. **Addressing Global Challenges**: The future of humanity will depend on our ability to address **global challenges** such as climate change, inequality, and political instability. These challenges require us to move beyond the limitations of our individual dimensions and work together to create **collective realities** that prioritize the well-being of all. By embracing dimensional awareness, we can build **bridges** between parallel realities and co-create solutions that transcend the boundaries of culture, ideology, and belief.

The Multiverse as a Model for Human Evolution

The Perceptual Multiverse is not just a theoretical framework; it is a model for **human evolution**. As we become more aware of the **multidimensional nature** of reality, we also become more aware of our **potential** as individuals and as a species. The multiverse offers a vision of humanity that is not limited by the constraints of **linear thinking** or **fixed beliefs**, but is instead characterized by **fluidity**, **creativity**, and **interconnectedness**.

1. **Dimensional Expansion**: As individuals and societies become more aware of the multiverse, we will continue to experience **dimensional expansion**. This involves moving beyond the limitations of traditional belief systems and exploring new ways of thinking, being, and interacting with the world. Dimensional expansion is a key part of

human evolution, as it allows us to adapt to new challenges, embrace new technologies, and expand our understanding of the universe.
2. **Conscious Co-Creation**: The multiverse also emphasizes the role of **conscious co-creation** in shaping reality. As we become more aware of the power of thought and perception, we can work together to co-create dimensions of reality that reflect our highest values and aspirations. This involves not only personal growth, but also collective action to build a more **equitable**, **sustainable**, and **compassionate** world.
3. **Exploring New Dimensions**: Finally, the Perceptual Multiverse invites us to explore new dimensions of reality, both within and beyond the physical world. Whether through **scientific discovery**, **spiritual practice**, or **creative expression**, humanity has the potential to unlock new dimensions of thought and existence that have yet to be fully understood. The future of human evolution will be marked by our ability to **transcend** the limitations of our current reality and explore the infinite possibilities that the multiverse offers.

Humanity's Journey into the Future

The **future of humanity** in the Perceptual Multiverse is one of **infinite potential**. As we continue to evolve, we will expand our awareness of the multiverse, exploring new dimensions of thought, belief, and existence. This journey will require us to embrace **dimensional fluidity**, cultivate **empathy**, and work together to co-create a future that reflects our highest values.

By recognizing that reality is not fixed, but is instead a dynamic and multidimensional experience, we open ourselves to the possibility of **profound transformation**—both on a personal and collective level. The Perceptual Multiverse offers a vision of humanity that is not bound by the limitations of the present, but is instead oriented toward the **expansion of consciousness**, the **exploration of new realities**, and the **co-creation of a better world**.

As we move forward into this multiverse of infinite dimensions, we carry with us the knowledge that our thoughts, beliefs, and perceptions shape the reality we experience. The future is not set in stone—it is a vast, multidimensional landscape waiting to be explored, understood, and co-created by all who inhabit it.

Ending and Gratitude

As we reach the conclusion of this exploration into the **Owens Perceptual Multiverse Theory**, I want to extend my deepest gratitude to you, the reader, for embarking on this journey with me. The concepts within this book—reality, perception, dimensions of thought—are complex and multifaceted, and I sincerely appreciate your time, curiosity, and willingness to dive deep into these ideas. My hope is that this theory has opened your mind to new possibilities, challenged your understanding of reality, and inspired you to see the world, and yourself, in a new light.

Whether this book has expanded your perception of the **multiverse of thought**, clarified the nature of **flashpoint issues**, or offered you a framework for personal or societal growth, I believe the potential for these ideas to shape our collective future is limitless. We are, each of us, co-creators of the reality we inhabit, and by embracing **dimensional awareness**, we have the power to navigate these infinite realities with intention, empathy, and wisdom.

As we conclude this book, I want to encourage you to continue exploring the dimensions of reality that resonate with you. Remember, the multiverse is not some distant, abstract concept—it is **here**, in every moment, shaped by your thoughts, beliefs, and perceptions. Every choice you make, every belief you challenge, and every new experience you embrace brings you closer to understanding the vast **perceptual universe** we inhabit.

This is not the end of our journey together. Stay tuned for my next book, **"How to Navigate the Perceivable Universes"**, where we will delve even further into the **practical applications** of the Perceptual Multiverse. In that book, we will explore **tangible techniques** for moving between dimensions, navigating conflicting realities, and using **dimensional awareness** to thrive in an ever-changing world. Together, we will learn how to harness the **power of perception** to navigate the complexities of life, relationships, and global challenges with grace and intention.

Thank you once again for your trust, your curiosity, and your commitment to exploring the vastness of the **multiverse of thought**. I look forward to continuing this journey with you in the next chapter of our exploration.

Until then, may your journey through the dimensions be one of growth, compassion, and endless discovery.

Warmest regards,
Anthony Scott Owens